Low-Grade Heat Harvesting

Low-Grade Heat Harvesting

Materials, Devices, and Technologies

Xiaogang Zhang

WILEY-VCH

Author

Prof. Xiaogang Zhang
Nanjing University of Aeronautics and Astronautics
Nanjing
China

Cover Image: © Xiaogang Zhang, Nanjing University of Aeronautics and Astronautics

Library of Congress Card No.: applied for

British Library Cataloguing-in-Publication Data
A catalogue record for this book is available from the British Library.

Bibliographic information published by the Deutsche Nationalbibliothek
The Deutsche Nationalbibliothek lists this publication in the Deutsche Nationalbibliografie; detailed bibliographic data are available on the Internet at <http://dnb.d-nb.de>.

Print ISBN: 978-3-527-35263-0
ePDF ISBN: 978-3-527-84366-4
ePub ISBN: 978-3-527-84367-1
oBook ISBN: 978-3-527-84368-8

Typesetting: Straive, Chennai, India
Printing and Binding: CPI Group (UK) Ltd, Croydon, CR0 4YY

C9783527352630_211223

Contents

About the Author

Prof. Xiaogang Zhang is professor of chemistry in the College of Material Science and Technology at Nanjing University of Aeronautics and Astronautics, Fellow of the Royal Society of Chemistry (FRSC), and the director of Jiangsu Key Laboratory of Electrochemical Energy Storage Technologies. He received his PhD degree in chemistry from Lanzhou University in 2001. His current research interests focus on the development of functional materials and devices in energy storage and conversion applications, including supercapacitors, batteries, ionic thermoelectrochemical systems, and sustainable energy. With a current H-index of 101, he was continuously identified in the global list of The Highly Cited Researchers (2016–2022).

Preface

Low-grade heat (<100 °C), as a typical and ubiquitous wasted energy, exhibits widely in nature, industrial activity, and our daily life. Unfortunately, tremendous amount of heat energy is still not harvested efficiently due to the lack of effective and sustainable technologies. Converting low-grade heat into electricity by thermoelectrochemical devices shows great potential for mitigating the energy crisis. However, there are great challenges to simultaneously achieve high power density and energy conversion efficiency as well as the integration with energy storage devices, which has greatly deferred the large-scale production of thermoelectrochemical devices and has become a key issue.

To accelerate the application of thermoelectrochemical systems, this book titled *Low-Grade Heat Harvesting: Materials, Devices, and Technologies* aims to discuss the cutting-edge progress for thermoelectrochemical systems from mechanisms to devices. In detail, this book covers topics including backgrounds and principles of low-grade heat harvesting, electronic thermoelectric devices for low-grade heat harvesting, polymer-based thermoelectric devices for low-grade heat harvesting, liquid-based thermocells for heat-to-current conversion, thermosensitive thermocells for low-grade heat harvesting, wearable power generation via thermoelectrochemical devices, thermoelectric ionogel for low-grade heat harvesting, alkali metal thermal electrochemical converter, thermally regenerative electrochemical cycle and other techniques, and integration of energy conversion and storage devices. As a rapidly growing frontier field in materials science and energy conversion, the topics involved in this book will generate significant attention from a variety of research fields and provide guidelines for those who are entering into a related research area or field. Finally, we sincerely appreciate all the contributors to this book.

Nanjing, China

Xiaogang Zhang

1

Backgrounds and Principles of Low-Grade Heat Harvesting

Developing sustainable energy is one of the most promising ways to mitigate the energy crisis caused by fossil fuels and greenhouse gas emissions. Low-grade heat (<100 °C), as a typical wasted energy, exhibits widely in nature, industry, and daily life. Except for geothermal energy, almost all generated heat resources (solar heat, machine heat, body heat, etc.) depend on some factors and cannot meet the continuous energy demand of the modern Internet of Things (IoTs) [1]. In particular, partial solar energy can be converted into electricity by solar cells based on the photovoltaic effect. However, the enormous amount of thermal energy is still not harvested efficiently. Therefore, the development of devices to convert heat into electricity should be the top priority at present.

Among the various technologies, solid-state thermoelectric (s-TE) and liquid thermocell (LTC) are the two dominant energy conversion technologies for high-value-added utilization of low-grade heat [2]. It is worth noting that all thermodiffusion-based cells (TDCs), thermogalvanic effect-based cells (TGCs), and thermoextraction-based cells (TECs) using liquid electrolytes fall into the category of LTCs. With the rapid development of sustainable energy, high power, and energy electronics, new and urgent demands have been placed on energy conversion technologies, such as the integration of TE conversion and storage. Unfortunately, most of the reported devices can only realize the energy conversion from heat to electricity, and external energy storage devices (i.e. capacitors and batteries) are required to store the charge generated from thermoelectrical processes, which increases the cost and complexity of the developed systems. In general, both capacitors and batteries are the leading energy storage technologies. Especially, Li-ion batteries as one of commercial devices are widely used in consumer products due to their high energy supply. In addition, Li-ion batteries have recently been developed to further improve the energy density of batteries and meet certain requirements. However, the sluggish kinetics of electron and ion during the charging and discharging processes may lead to partial energy loss, which increases heat generation and dendrite formation. In addition, some reported failures in electric cars, airplanes, and energy storage power plants prompt us to pay attention to battery thermal management. Electrochemical capacitors, also known as

Low-Grade Heat Harvesting: Materials, Devices, and Technologies, First Edition. Xiaogang Zhang.

supercapacitors, can output a high power density and complement batteries in some areas due to their low cost, long cycle life, and satisfactory safety. However, many efforts are being made to improve the relatively low energy density of capacitors to meet the growing interest in high energy and power. Therefore, the development of thermoelectrochemical devices with integrated energy conversion and storage, which have fast response, long durability, and high energy/power density, is of great importance [3]. To our knowledge, TE devices can be classified into three different forms, including organic Rankine cycles, traditional TEs, and thermocells. TEs can be further divided into two types based on the electrode materials used: inorganic TEs and organic TEs. Thermocells can be identified as devices based on electric double-layer capacitive (EDLC) mechanism and pseudocapacitive behavior.

1.1 Backgrounds and History

The development history of thermoelectrical devices is a story of discovering energy conversion mechanisms. The demonstration of direct conversion from heat to electricity can date back to 1822. The first physical effect, named "Seebeck effect," was found by German scientist, Thomas Johann Seebeck. As displayed by the timeline in Figure 1.1, two different metal wires were connected to form a current loop in this experiment. When heating one of the junctions while the other junction was maintained cold, there was a magnetic field exhibiting around the circuit. Although Seebeck did not provide a correct explanation for this interesting phenomenon, it did not prevent him from conducting comparative works on many materials, which laid the foundation for later thermoelectrical studies. On the basis of this design, J. C. A. Peltier found that the temperature near the junction can be changed when current flowed through two different metals in 1834, which is also called as "Peltier effect", confirming the TE effect. Until 1838, the feature of the Peltier effect was rationally explained by Heinrich Friedrich Emil Lenz: whether the junction of two different conductors absorbs or releases heat depends on the direction of current flowing through the circuit, and the amount of heating (cooling) is proportional to the magnitude of the current. As a proof of concept, the preformed ice on the junction can be melted into water only by changing the direction of the current.

Both the Seebeck effect and the Peltier effect are discovered on the junction of two different conductors; however, those are still not interface effects. The relationship between the TE effect was not well recognized until the nineteenth century. In 1857, William Thomson (Lord Kelvin) made a comprehensive analysis of the Seebeck effect and the Peltier effect using the thermodynamic principle established by himself. Thomson believed that there was a simple multiple relationship between the Peltier coefficient and the Seebeck coefficient at absolute zero. On this basis, he theoretically predicted a new TE effect, that is, when a current flows through a conductor with uneven temperature, the conductor will not only generate irreversible Joule heat but also absorb or release a certain amount of heat. This new TE effect is also known as the Thomson effect.

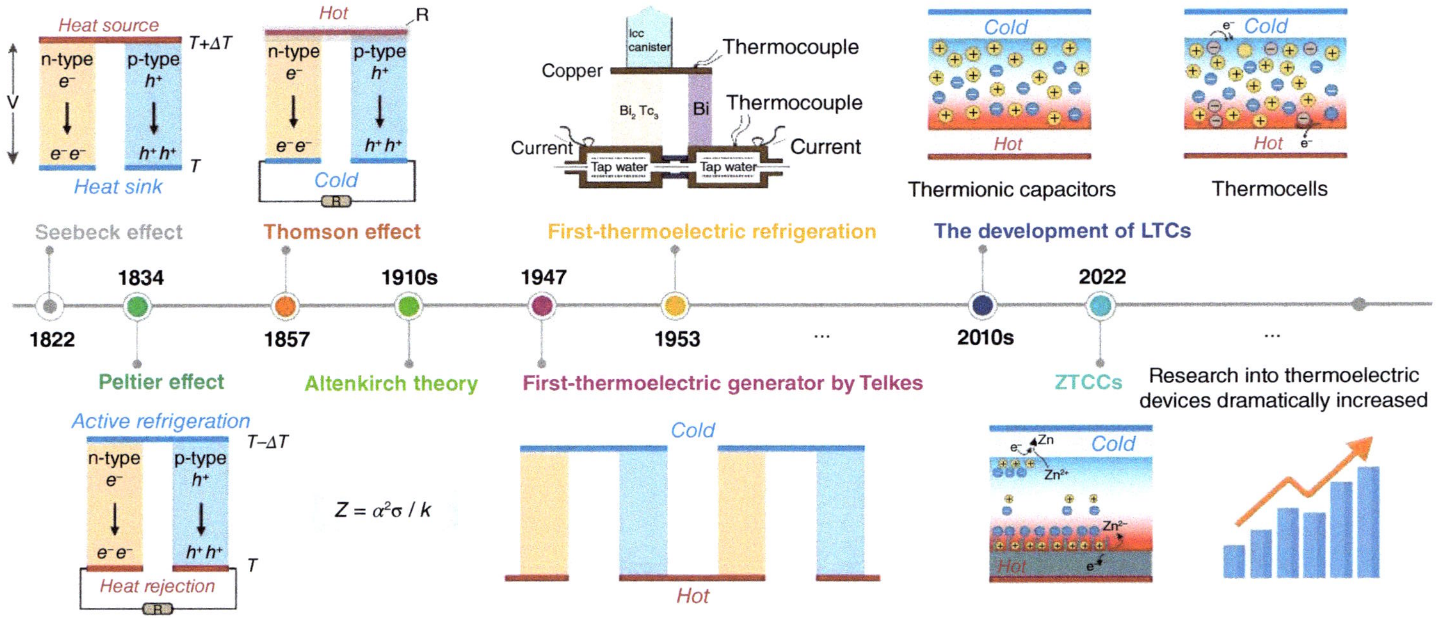

Figure 1.1 Historic timeline for the development of thermoelectric devices.

In the 1910s, Edmund Altenkirch proposed a satisfactory theory for TE refrigeration and power generation. In detail, a promising TE material must have the following merits including a large Seebeck coefficient (S, ensure relatively obvious TE effect), high electrical conductivity (σ, reduce the generated Joule heat), and low thermal conductivity (κ, retain heat near the junction). It is worth mentioning that the relationship among such parameters can be described as: $Z = S^2\sigma/\kappa$, which can be used to evaluate the thermoelectrical performances in practical applications. Since only metals and their compounds were considered important conductors at the time, researchers focused their attention on metals and corresponding alloys, ignoring semiconductor materials. Nevertheless, most metals exhibit a very low Seebeck coefficient ($\sim$10 $\mu V\,K^{-1}$), which not only results in a low conversion efficiency but also limits the development of thermoelectrical devices.

With the rapid development of semiconductor materials in the middle of the twentieth century, researchers found that the Seebeck coefficient of semiconductors was 10 times higher than that of metals. In 1947, the first semiconductor-based TE generator was invented by Telkes, which can deliver a heat-to-current efficiency of 5%. Meanwhile, the first TE refrigeration prototype was fabricated in 1953 by Bi_2Te_3 and Bi as working electrodes. However, the advantages of semiconductor as TE material are not fully manifested due to the significant difference in conductivity between semiconductors and metals in the time. In the following time, a lot of works are focused on the properties optimization of semiconductors to obtain a high energy conversion efficiency.

In the 2010s, a new class of thermoelectrochemical system, termed LTCs, which includes thermionic capacitors and pseudocapacitive thermocells, was proposed. Typically, the thermoelectrochemical performances of LTCs can be greatly improved by electrode development and electrolyte optimization. For example, Crispin et al. constructed an ionic TE supercapacitor through a remarkably strong ionic Soret effect (thermal diffusivity) using polymeric electrolyte and carbon nanotube (CNT) electrode, which can realize the conversion of heat into stored charge [4]. On the other hand, Liu and Chen et al. have proposed a new ionic TE material based on the synergistic TD and TG effects, demonstrating significant promise for heat-to-current conversion using ions as energy carriers and opening up a new research field for high-performance thermoelectrochemical devices [5]. In 2022, Zhang and co-workers demonstrated a zinc ion thermal charging cell (ZTCC) based on the TD and TE of electrolyte ions. Because of the unique feature of a multivalent charge carrier and the relatively low potential of zinc anode, a high output voltage and large ionic Seebeck coefficient/thermopower can be achieved [6].

Since the 2020s, the amount of research related to TE devices has continuously and dramatically increased in line with the emerging increased demand for sustainable, flexible, high performance, and safe energy conversion-storage devices. Moreover, many advanced materials and devices have been discovered for high-value-added conversion of low-grade heat into stored energy, along with the development of nanoscience and characterization techniques.

1.2 Working Principles of Current Systems

Typically, the emerging technologies for low-grade heat harvesting include TEs and thermoelectrochemical devices. For TEs, the involved mechanisms can be divided as: (i) Seebeck effect, (ii) Peltier effect, and (iii) Thomson effect.

1.2.1 Seebeck Effect

The characteristic of the Seebeck effect can be simply explained by the change of charge distribution in the conductor under temperature gradient. For p-type semiconductors, the material is electrically neutral due to the even distribution of internal carriers under a uniform temperature field. As shown in Figure 1.2, with adopting temperature difference on both ends of the conductor, a large amount of holes can be aggregated on the hot side due to the continuous energy input. Consequently, the uneven distribution of internal charge concentration leads to an electrical field or potential difference in materials. Meanwhile, under the action of the potential difference, a reverse drift charge flow could be generated. When the charge diffusion flow of thermal motion is equal to the drift charge flow generated by the internal electric field, a dynamic equilibrium is reached, and a stable thermal-induced potential (V) can be formed at both ends of the semiconductor. Accordingly, the Seebeck coefficient (S) of material at a specific temperature (T) can be defined as:

$$S = \lim_{\Delta T \to 0} \frac{V}{\Delta T} \tag{1.1}$$

Similarly, the direction of thermal-induced potential for n-type semiconductors is opposite to that of p-type materials.

Figure 1.2 Illustration of Seebeck effect using p-type materials.

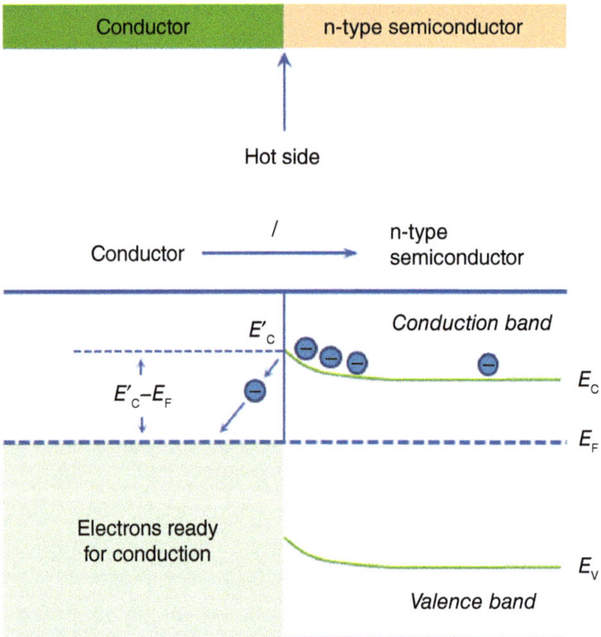

Figure 1.3 Illustration of the Peltier effect.

1.2.2 Peltier Effect

As displayed in Figure 1.3, when electrons flow from n-type semiconductor with high energy level to a conductor with low energy level under the action of an electric field, the electrons transition is downward at the interface barrier, resulting in a macroscopic exothermic process. When electrons flow from a conductor with low energy level to a conductor with a high energy level, they will absorb a certain amount of heat and jump upward, which is manifested as an endothermic effect. Moreover, the heat absorbed or released per unit time is proportional to the current intensity, as described by the following equation:

$$\frac{dQ}{dt} = \pi I \tag{1.2}$$

where π is the differential Peltier coefficient, t is the operation time, and I is the adopted current. Notably, when the current flows from the metal conductor to the p-type material, the system shows heat absorption, and the differential Peltier coefficient is negative. If the current is reversed, the Peltier coefficient also reverses positive and negative accordingly and has directionality.

1.2.3 Thomson Effect

The discovery of the Seebeck effect and the Peltier effect both involve a loop composed of two metals and both occur at the junction of different conductors, but neither of them is an interfacial reaction. Based on our current knowledge, we find that they both originate from the intrinsic properties of two conductors that make up the junction.

In fact, the current regarded Thomson effect was verified experimentally in 1876. When the temperature difference of the conductor along the current direction is ΔT, the heat released (or absorbed) per unit time on this conductor can be expressed as:

$$\frac{\mathrm{d}Q}{\mathrm{d}t} = \beta \Delta T I \tag{1.3}$$

where β represents the Thomson coefficient. When the current direction is consistent with the temperature gradient, if the conductor absorbs heat, the Thomson coefficient is positive; otherwise, it is negative. Compared with the Seebeck effect and Peltier effect, the Thomson effect contributes very little to the energy conversion in the TE conversion process, so it is often ignored in the design of TE devices and energy conversion analysis.

Thermoelectrochemical devices are another category of energy conversion systems based on the diffusion of electrolyte ions or Faradaic redox processes. Generally, the electricity is generated from the concentration difference between cation and anion caused by their different thermal mobility as well as the entropy change. Moreover, the standard electrode potential can be improved at the isothermal condition with the introduction of redox species. According to the background integrated transport property and thermodynamic behavior, various electrode materials and electrolyte systems have been developed to construct promising thermoelectrochemical devices. Although the combination of the TD and TG effects can achieve a high thermopower (or Seebeck coefficient), the relatively low output voltage and complex packaging still limit the practical applications of common thermoelectrochemical devices. Therefore, the proposal of satisfying thermoelectrochemical devices with large Seebeck coefficient, high efficiency, and low cost shows great promise to solve the bottleneck of current technology for high-value-added low-grade heat harvesting near room temperature.

Different from the traditional s-TEs, the LTCs mainly involve the electrochemical processes such as electrostatic adsorption/desorption and near-surface redox reactions, whereas TEs display the physical process dominated by directional transport of electrons and holes. It is necessary to clarify that the direct introduction of concepts about TEs into the LTCs is inappropriate. For the TD effect, the thermopower derived from ions can be defined as:

$$S_{\mathrm{td}} = -\frac{V(T_{\mathrm{H}}) - V(T_{\mathrm{C}})}{T_{\mathrm{H}} - T_{\mathrm{C}}} \tag{1.4}$$

where $V(T_{\mathrm{H}})$ and $V(T_{\mathrm{C}})$ represent the output voltage of electrode at hot temperature (T_{H}) and cold temperature (T_{C}), respectively. Notably, this concept is similar to conventional TEs because of their similar nature of carrier ions mobility. The essential driving force of LTCs depends on the transport of entropy from the hot end to the cold end, which is mainly contributed by the TG effect. Specifically, the redox couples in electrolyte are typically oxidized at the electrode on hot side ($\mathrm{O} + n\mathrm{e}^- \rightleftharpoons \mathrm{R}$), while reduction reaction occurs on cold side. The reaction products are brought to the original place by the convection in electrolyte, and thus generate thermo-voltage under a certain temperature gradient (ΔT), which is defined as temperature coefficient (α), expressed as:

$$\alpha = \frac{\partial V}{\partial T} = -\frac{1}{nF}\frac{\partial \Delta G}{\partial T} = \frac{\Delta S}{nF} \tag{1.5}$$

where ΔG, ΔS, F, and n are the Gibbs free energy, entropy change, Faraday's constant, and the number of electrons transferred in the reaction, respectively. Accordingly, the increase in concentration difference and the entropy change under low-temperature gradient is an efficient way to enhance the performance of the TD-dominated device and TG effect-governed cells, respectively.

Nowadays, several works about LTCs have been contributed with the aim of achieving high-performance conversion of low-grade heat together with the design of functional electrolytes. It should point out that the development of electrode with high electrical conductivity, large surface area, and catalytic properties also can boost the output performances of LTCs. Under this consideration, well understanding of the energy conversion modes and the mechanisms between functional electrodes and electrolytes is critical to build satisfactory LTCs.

In detail, the energy conversion mechanism of thermoelectrochemical devices can mainly be classified into three types: (i) TD effect, (ii) TG effect, and (iii) TE effect.

1.2.4 Thermodiffusion Effect

Generally, the mobility of electrolyte ions is the most important property of TD processes. As shown in Figure 1.4a, the cations with relatively small size can diffuse from hot side to cold side under a temperature gradient. It is worth mentioning that the anions with relatively large size would also move with the continuous input of heat. For convenience, we regraded that only cation will move and the anion will be accumulated on the hot side during the TD process. Consequently, a reasonable concentration difference of cations can be formed by the separation of electrolyte ions, which can lead to a voltage output. The potential generated from TD can be typically estimated according to the following equation [7]:

$$E_{\mathrm{TD}} = E^{\ominus}_{\mathrm{TD}} - \frac{RT}{nF}\ln\frac{[\mathrm{Cation}]_{\mathrm{H}}}{[\mathrm{Cation}]_{\mathrm{C}}} \tag{1.6}$$

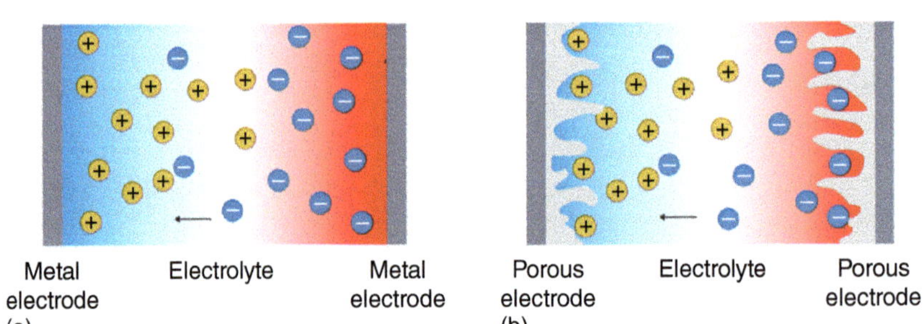

Figure 1.4 Illustration of thermodiffusion effect based on (a) metal electrode and (b) porous electrode.

where E_{TD} and $E_{TD}^{\ominus}$ represent the potential difference and the standard potential between two electrodes, $[\text{Cation}]_H$ and $[\text{Cation}]_C$ are the concentration of cations near the hot and cold side, R, T, and F refer to the ideal gas constant (8.314 J mol^{-1} K^{-1}), temperature, and Faraday constant (96485.338 C mol^{-1}), respectively. Inspired by this principle, the size of electrolyte ions should be seriously considered to obtain a high cation concentration difference between the two sides. In addition, the TD processes only involve the adsorption and desorption of ions on hot/cold side. Thus, the performances (i.e. Seebeck coefficient, output voltage) of TD-dominated cells (TDCs) with metal electrodes are always at a low level.

As an efficient strategy, the TDCs can achieve attractive heat-to-current performances by replacing the metal electrodes with porous electrodes (i.e. active carbon, carbon nanotubes, and graphene) [8]. Benefitting from the structural advantages of porous electrodes including large surface area, abundant porosity, high electrical conductivity, and good stability, more electrolyte ions can be stored in porous electrodes than in plane metal electrodes, significantly enhancing the thermoelectrochemical behaviors of TDCs (Figure 1.4b). In fact, the TD procedure can belong to the category of EDLC behaviors. The open-circuit voltage (OCV) for TDC shows a linear relationship with the change of time, further confirming the EDLC processes. Therefore, the interaction between electrolyte and electrode should be well understood for further construction of TDCs.

1.2.5 Thermogalvanic Effect

For TGCs, the electrolyte containing redox species is sandwiched between two electrodes with different temperatures (Figure 1.5a). Due to the reversible reactions of redox couples, the entropy change can be found in TGCs during reduction and oxidation, therefore generating a thermovoltage. Notably, the electron associated with the redox reactions can move from the oxidized side to the reduced side through the external circuit. According to the Nernst equation, the equilibrium potential (E_{TG}) of TGC can be determined by the following equation [9]:

$$E_{TG} = E_{TG}^{\ominus} - \frac{RT}{nF} \ln \frac{[\alpha_R]_H}{[\alpha_O]_C} \tag{1.7}$$

where $E_{TG}^{\ominus}$ is the standard potential, and α_R and α_O represent the activities of reduced species and oxidized species, respectively. Moreover, the activity can be defined as the product of activity coefficient (γ_i) and concentration (C_i):

$$\alpha_i = \gamma_i \times C_i \tag{1.8}$$

Thus, Eq. (1.5) can be described as:

$$E_{TG} = E_{TG}^{\ominus} - \frac{RT}{nF} \left[\ln \frac{[\gamma_R]_H}{[\gamma_O]_C} + \ln \frac{[C_R]_H}{[C_O]_C} \right] \tag{1.9}$$

In typical TGCs with metal electrodes, both concentrations of reduced species and oxidized species at hot side and cold side are same, that is, the output potential only depends on the activity coefficient of both redox couples.

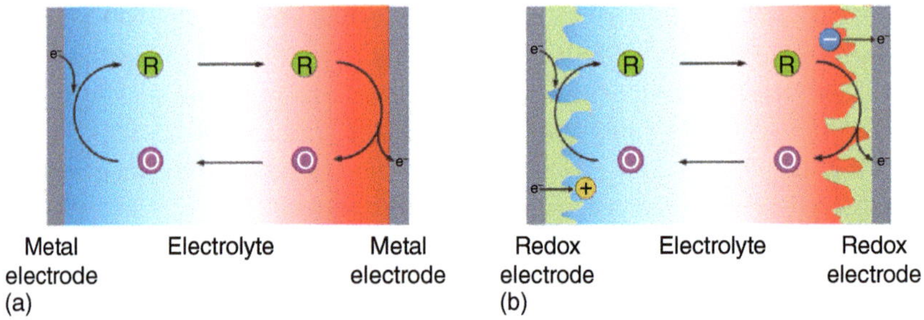

Figure 1.5 Illustration of thermogalvanic effect based on (a) metal electrode and (b) redox electrode.

In contrast to traditional TGCs, the thermoelectrochemical performances can be further improved by changing the metal electrodes with pseudocapacitive electrodes. As known, the charge transfer (Faradaic process) at the surface or near the surface of pseudocapacitive materials exhibits fast and reversible features. As displayed in Figure 1.5b, the redox reactions that occurred between the ion carriers and active electrode materials would greatly enhance the output voltage and the thermopower of modified TGCs, owing to the integration of typical TG effect, redox-free ions concentration gradient, and contribution from redox electrode materials. Accordingly, the OCV curves for TGCs display visible platform attributed to the related redox reactions.

1.2.6 Thermoextraction Effect

As one attractive device among the most low-grade heat harvesting systems, TECs combine the advanced properties of both electrode and electrolyte. Most importantly, as-constructed TECs can realize the integration of energy conversion and storage. As illustrated in Figure 1.6a, the TECs are assembled by the insert-type materials (i.e. $LiMn_2O_4$), EDLC materials (i.e. porous carbon), and functional electrolytes [10]. Typically, the cations inserted in positive electrodes can be extracted and transported to negative electrodes under temperature gradient. During this process, the electron produced from the redox reaction of cations extraction can move from hot side to cold side through the external circuit. After the

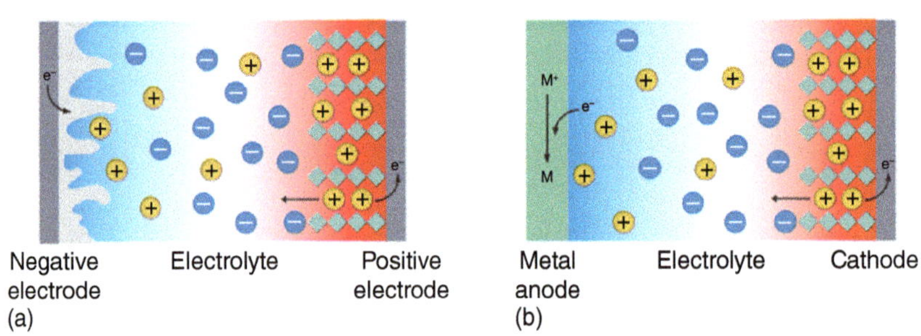

Figure 1.6 Illustration of TE effect based on (a) capacitor and (b) battery mechanism.

charge accumulates on the surface of negative electrode, a plenty of cations can be adsorbed, which leads to a high concentration difference of ions between electrodes as well as the thermovoltage. Thus, the TE processes still belong to the category of pseudocapacitive behavior-induced thermoelectrochemical conversion. Similar to the above-mentioned strategies for the development of high-performance devices, both porous electrodes and insert-type electrodes should possess abundant electroactive sites, good conductivity, and high capacity, providing enough capability for ions storage and charge transfer.

Electrochemical metal anodes have recently emerged as the most promising electrode to build batteries or capacitors for energy storage due to the unique properties of metal anodes. When integrating the facile metal anode like zinc, a water-compatible device can be proposed for energy conversion and storage [6]. As shown in Figure 1.6b, the cations inserted in cathode materials can be gradually extracted with the heat input. After moving to anode side, such cations get deposited on the surface of metal anode. The formed electrons transfer from hot side to cold side during the whole thermoelectrochemical process through the external circuit. Thanks to the multivalent feature of Zn^{2+}, a high energy density and heat conversion efficiency can be obtained by TECs. Notably, attractive OCV curves can be obtained for TECs, and obvious redox feature together with relatively high kinetics can greatly enhance the output voltage. Excitingly, the metal anode can be replaced with other promising candidates, such as lithium, sodium, potassium, and some pre-metalized anodes. Worthily, the compatible electrolyte engineering and device integration should be well considered for the development of nonaqueous TECs to avoid potential safety problems.

Besides, the use of redox-free porous materials as electrodes for TGCs can also improve the output performances of devices by the combination of TD and TG effects. When employing redox-free materials as anode and redox materials as cathode, another TGC can be developed similar to "asymmetric capacitor." We believe that more ideal mechanisms and devices will be proposed with the development of technology for energy conversion and storage.

1.3 Parameters for Low-Grade Heat Harvesting

The unitless TE figure of merit (ZT) of materials is an important criterion for judging the TE properties of materials. When taking n-type semiconductor as an example, the electrons and holes can be regarded as their major and minor carriers. Therefore, the relationship of S, σ, and κ can be expressed as follows [11]:

$$S = \frac{8\pi^2 {k_B}^2}{3eh^2} m^* T \left(\frac{\pi}{3n}\right)^{2/3} \tag{1.10}$$

where κ_B, e, h, and m^* represent the Boltzmann constant, the electrical charge, the Planck constant, and the carrier effective mass, respectively.

$$\sigma = ne\mu \tag{1.11}$$

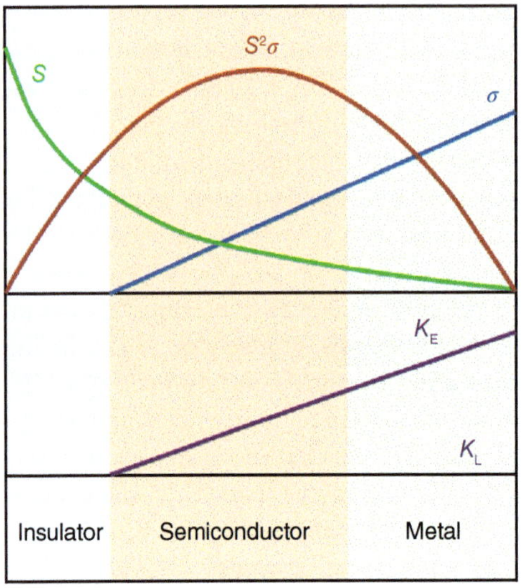

Figure 1.7 The relationship among S, σ, and κ for different materials.

where μ represents the carrier mobility.

$$\kappa = D_T C_P \rho = \kappa_e + \kappa_1 \tag{1.12}$$

$$\kappa_e = L\sigma T = ne\mu LT \tag{1.13}$$

where D_T, C_P, ρ, and L represent the thermal diffusivity, the specific heat, density, and the Lorenz number, respectively.

For a long time, research on TE materials focused mainly on improving the ZT value. According to $ZT = S^2\sigma/\kappa$, high-performance TE materials must have a high Seebeck coefficient, high electrical conductivity, and low thermal conductivity. Based on these relationships, the values of S, σ, and κ for electrical materials should be balanced to optimize a high ZT value. In Figure 1.7, the relationship between the three parameters affecting the TE properties of materials and the carrier concentration is shown qualitatively. The Seebeck coefficient and the conductivity of the material are related to the charge carrier concentration of the material. As the carrier concentration increases, the conductivity increases and the Seebeck coefficient decreases, with the changes in the two parameters showing an opposite trend. For most semiconductor materials, the magnitude of the optimum carrier concentration with the best TE properties is usually 10^{19}–10^{20} cm^{-3}, which corresponds to a heavily doped state (or degenerate state). The thermal conductivity can usually be divided into the thermal conductivity of the charge carriers and the thermal conductivity of the lattice. The thermal conductivity of the charge carriers is proportional to the electrical conductivity. Therefore, too high electrical conductivity directly leads to high thermal conductivity, which is not conducive to improving the ZT value of the material. The thermal conductivity of the lattice is relatively independent. However, since the scattering mechanisms of electrons and thermal conduction phonons are usually interconnected and interact, a decrease in lattice thermal conductivity due to scattering phonons often affects the electrical conductivity and Seebeck coefficient. Therefore, the three important parameters, S, σ, and κ, are closely related to each other and significantly determine the TE properties of materials. Increasing or decreasing one of the parameters alone often leads to non-synergistic changes in the other parameters. For this reason, it is difficult to continuously improve the TE performance ZT. Therefore, realizing independent or coordinated control of current and heat transport is a goal that TE materials science has long pursued.

S, σ, and κ are the main parameters that determine the TE properties of materials, and the accurate measurement of these three parameters is the core content of the performance characterization of TE materials. The Seebeck effect and the Peltier effect often interfere with the accurate measurement of the electrical and thermal properties of TE materials. After many years of technical improvements, the measurement technique for TE transport properties of bulk materials is relatively mature, but the complexity of measurement error sources and incomplete uniformity of measurement standards are still difficult problems for TE materials researchers. Here, we will focus on the measurement principles and methods of Seebeck coefficient, electrical conductivity and thermal conductivity, analyze the causes of measurement errors, and search for solutions.

1.3.1 Seebeck Coefficient

In the s-TE field, the Seebeck coefficient is the key instinct parameter of materials. As defined in Eq. (1.1), the S value highly depends on the adopted temperature difference (ΔT) and as-generated potential difference (ΔV). In a typical run, a temperature gradient can be established on the sample for test through the heating plate and cooling plate at the upper and lower ends. Meanwhile, the potential difference between two sides can be determined by the voltage probe. Accordingly, some conditions should be considered, including (i) stable and uniform temperature and voltage, (ii) temperature and voltage tests at the same position, and (iii) only temperature-induced case. Based on these principles, a series of ΔT and corresponding ΔV values can be obtained. When the ΔT is small enough, the Seebeck coefficient can be fitted as the slope of the function of ΔV and ΔT. It is worth mentioning that such a proposed method is the most used way for measuring the Seebeck coefficient of TE materials.

1.3.2 Electrical Conductivity

Electrical conductivity is one of the basic electrochemical properties of functional materials. Compared with the measurement of the Seebeck coefficient, the testing method of electrical conductivity is relatively simple and mature. For a material with a uniform composition, the electric field inside the material is evenly distributed when a constant current is applied.

At this time, the electric field strength inside the material can be described as:

$$E = V/l \tag{1.14}$$

and the current density flowing through the material can be written as:

$$J = I/A \tag{1.15}$$

where l and V are the distance of two sides and corresponding potential differences, and A and I are the cross-sectional area and current.

Thus, the electrical conductivity (σ) of materials can be determined by:

$$\sigma = \frac{J}{E} = \frac{Il}{VA} \tag{1.16}$$

Typically, the measurement methods of material conductivity mainly include the two-probe method and the four-probe method. Taking the four-probe method as an example, the sample for measurement should have a uniform composition, and the probes and the surface of sample should be kept in contact as much as possible. Besides, some other factors also need to be considered in actual measurement. To our knowledge, the thermoelectrical materials like semiconductors exhibit low resistivity, which would lead to a non-negligible error. As the self-oxidation of most materials such as alloy in atmosphere, as-formed oxide layer on the surface of materials causes a relatively large contact resistance between sample and metal probe, which will generate additional nonohmic voltage and cause test errors. To eliminate the influence of contact resistance, the slope of I–V curve recorded by various voltage signals under different currents is generally employed to measure the real resistance of testing sample. Notably, TE material always has a large Seebeck coefficient and Peltier coefficient, which may inevitably form a temperature difference ΔT at both ends of the testing sample. This process further generates additional Seebeck potential during actual test, affecting the accuracy of obtained results. Besides, the applied current should not be too high to reduce the temperature fluctuation of the sample caused by self-generated Joule heat.

1.3.3 Thermal Conductivity

Thermal conductivity reflects the heat conduction ability of testing materials. According to Fourier's law, thermal conductivity can be defined as the heat transferred by a unit temperature gradient (the temperature drops by 1 K within a length of 1 m) through a unit heat transfer surface in a unit time. However, due to the existence of various forms of heat exchange such as radiation, conduction, and convection, the key to accurately measuring thermal conductivity is to solve the problem of heat exchange between the testing sample and environment. It is one of the most important goals to achieve thermal insulation.

In general, the methods used for measuring thermal conductivity of samples mainly include steady-state method and nonsteady-state method. The steady-state method is the earliest method used in thermal conductivity measurement. The sample to be tested is placed between the heater and the radiator, and a stable heat source is applied at one end to keep it in a steady state. By measuring the temperature difference between the two ends of samples and the corresponding heat flux flowing, the thermal conductivity of materials can be calculated by the following equation:

$$J_{\mathrm{T}} = -\kappa \frac{\mathrm{d}T}{\mathrm{d}x} \tag{1.17}$$

where κ represents the thermal conductivity. $\mathrm{d}T/\mathrm{d}x$ and J_{T} are temperature gradient and heat flux density, respectively.

It should be pointed out that the J_{T} can be directly obtained by the power of heater applied during test. However, the thermal resistance and heat exchange will cause large errors. Some great efforts have been contributed to solve this problem.

Nonsteady-state method is a rapid measurement method developed to solve the problems of long measurement time and great influence of heat loss on measurement accuracy in the steady-state heat flow method. According to the different ways of applying heat source, nonsteady-state methods mainly include periodic heat flow method and transient heat flow method. The basic principle is to apply periodic heat flow or transient (pulse) heat flow on the sample, and then measure the temperature change of sample to calculate thermal conductivity. The laser pulse method is a transient heat flow method developed in the 1960s and has become one of the most commonly used and mature thermal conductivity measurement methods [12]. In this way, the thermal conductivity of the sample can be calculated by applying a pulsed laser to heat one side, and then measuring the temperature change on the other side caused by heat conduction.

Abovementioned methods are typically employed in block thermoelectrical materials. Moreover, the accuracy of test is greatly affected by the shape and size of samples. Thus, some advanced methods should be developed for the characterization of other systems.

1.3.4 Conversion Efficiency

Among the state-of-the-art techniques, TDCs and thermocells are regarded as potential devices besides s-TEs for the direct conversion of low-grade heat into electricity. When evaluating the thermoelectrochemical performances of such devices, the ratio of device voltage output to temperature difference input ($\Delta V/\Delta T$) is always used as a key numeric performance index to characterize the energy conversion ability of devices. However, some differences still should be claimed.

Typically, the thermoelectrochemical performances can be evaluated by some parameters including ionic Seebeck coefficient (S), power density (P), normalized power density ($P/(A\ \Delta T^2)$), and conversion efficiency (η). From previous reports, various methods were used to calculate these values. However, the applicable conditions of different methods should be carefully considered to accurately report the results of developed systems [13]. Establishing the standardized principle for as-obtained performance is very helpful for researchers in thermoelectrochemical areas to avoid the confusion.

For as-assembled thermoelectrochemical devices, the energy conversion efficiency (η) can be defined as the ratio of the maximum electrical power output (P_{max})

to the total heat power input (P_{heat}) [14]:

$$\eta = \frac{P_{max}}{P_{heat}} \tag{1.18}$$

In a typical run, P_{heat} represents the heat flux through the cell, which can be obtained by the following equation:

$$P_{heat} = \kappa_{eff} A \frac{\Delta T}{d} \tag{1.19}$$

where κ_{eff} is the effective thermal conductivity, A is the cross-sectional area, ΔT is the temperature difference between hot side and cold side, and d is the distance of both electrodes.

Under this case, the energy conversion efficiency can be written as:

$$\eta = \frac{P_{max}}{P_{heat}} = \frac{P_{max}}{\kappa_{eff} A(\Delta T/d)} \tag{1.20}$$

When considering the limitation of Carnot efficiency ($\eta_c = \Delta T/T_{hot}$) for a heat engine, the Carnot-related efficiency (η_r) is employed as one important criterion to evaluate the practicability of as-constructed devices, which can be further defined as:

$$\eta_r = \frac{\eta}{\eta_c} = \frac{P_{max} / \left(\kappa_{eff} A \frac{\Delta T}{d}\right)}{\frac{\Delta T}{T_{hot}}} = \frac{P_{max}\, dT_{hot}}{\kappa_{eff} A \Delta T^2} \tag{1.21}$$

Here, it is worth mentioning that the P_{max} can be obtained from the relationship between OCV (V_{oc}) and short-circuit current (I_{sc}):

$$P_{max} = \frac{V_{oc} I_{sc}}{4} \tag{1.22}$$

Thus, the Carnot-related efficiency can be expressed as:

$$\eta_r = \frac{V_{oc} I_{sc} \mathrm{d} T_{hot}}{4\kappa_{eff} A \Delta T^2} \tag{1.23}$$

For most reported thermoelectrochemical systems, the relationship between V_{oc}, $I_{sc,}$ and ionic Seebeck coefficient can be defined as $V_{oc} = \alpha\Delta T$ and $I_{sc} = \alpha\Delta T/R_{cell}$, where R_{cell} is the internal resistance of systems. Thus, the following equations are obtained:

$$\eta = \frac{\alpha^2 \mathrm{d} \Delta T}{4\kappa_{eff} A R_{cell}} \tag{1.24}$$

$$\eta_r = \frac{\alpha^2 \mathrm{d} T_{hot}}{4\kappa_{eff} A R_{cell}} \tag{1.25}$$

Here, the measurement of thermal conductivity should be seriously emphasized. Due to the high mobility of liquid, the thermal conductivity could change in a large range when applying a temperature or force disturbance. Generally, the thermal conductivity of liquid can be derived from thermal conduction and thermal convection. Typical methods like hot-wire method and laser flash method are performed to measure the intrinsic thermal conductivity of materials under an isothermal and non-flow state, in which the thermal convection can be negligible. However, these methods are not applicable for measuring the effective thermal conductivity of liquid in thermocells under a non-isothermal state. Because of the existence of large temperature difference in thermocells, the considerable convection formed in system can significantly enhance the thermal conductivity of applied liquid. Alternatively, the steady-state method based on the heat flux conservation law can directly measure the effective thermal conductivity from both thermal conduction and convection. As discussed above, the energy conversion efficiency of thermoelectrochemical devices is highly dependent on their effective thermal conductivity. Thus, the measurement of thermal conductivity for LTCs should be well considered to avoid the overevaluation of thermoelectrochemical performances.

Besides, another method for the calculation of energy conversion efficiency is established in heat recuperation. As described, the heat-to-current efficiency can be determined by the quotient of output electrical energy (W) and input heat energy (Q). Such general expression can be written as [15]:

$$\eta = \frac{W}{Q} = \frac{W}{Q_{\mathrm{H}} + Q_{\mathrm{dis}}} \tag{1.26}$$

where Q_{H} and Q_{dis} are the adsorbed heat and the continuous energy input at T_{H}, respectively.

Owing to the heating from T_{C} to T_{H}, the heat consumption involved in this process can be calculated by using this equation:

$$Q_{\mathrm{H}} = (1 - \eta_{\mathrm{HX}}) \sum m_i c_{p,i} \Delta T \tag{1.27}$$

where m_i and $c_{p,i}$ represent the mass loading and specific heat capacity of the component i (such as anode, cathode, electrolyte, and current collector). η_{HX} is the efficiency of heat recuperation. ΔT is the temperature difference applied, which can be determined by $T_{\mathrm{H}} - T_{\mathrm{C}}$.

Meanwhile, the energy consumption associated with chemical reactions requires continuous heat input to maintain isothermal conditions during discharge states, which can be defined as:

$$Q_{\mathrm{dis}} = T_{\mathrm{H}} \Delta S_{\mathrm{H}} \tag{1.28}$$

where ΔS_{H} is the total entropy change, which equals to the entropy change at cathode ($\Delta S_{\mathrm{H+}}$) and anode ($\Delta S_{\mathrm{H-}}$).

Based on the partial molar entropy change (Δs_i) and the amount of component (i), the ΔS_{H} can be expressed as:

$$\Delta S_{\mathrm{H}} = \int \Delta s_i \mathrm{d}n_i \tag{1.29}$$

For electrochemical reactions involved during heat recuperation,

$$\frac{\mathrm{d}\Delta G}{\mathrm{d}\Delta T} = -\Delta S \tag{1.30}$$

$$\Delta G = -nFE \tag{1.31}$$

Therefore,

$$\Delta S_H = \int nF\frac{\Delta E}{\Delta T}\mathrm{d}n_i = \alpha \int nF\mathrm{d}n_i = \alpha q_{\mathrm{dis}} \tag{1.32}$$

where q_{dis} is the charge transferred during discharge.

When applying Eq. (1.32) to (1.28),

$$Q_{\mathrm{dis}} = T_{\mathrm{H}}\alpha q_{\mathrm{dis}} \tag{1.33}$$

It is worth mentioning that the specific capacitance C (F g^{-1}) according to the report from F. Béguin's group for GCD process [16]:

$$C = \frac{2 \times I \times S}{m \times \Delta V^2} = \frac{2 \times I \int_{t_0}^{t_1} V(t)\mathrm{d}t}{m \times \Delta V^2} \tag{1.34}$$

where S (Vs) is the integral area of discharge curve. I (A), m (g), ΔV (V), and t (s) represent the current, the mass of active materials, the voltage range operated, and discharging time. After unit conversion and normalization of the active mass loading, the q_{dis} can be calculated.

Besides, the total network output in this experiment is measured by galvanostatic discharge technique and can be calculated from:

$$W = \int_{t0}^{t1} (V_1 - V_0) I_{\mathrm{dis}}\mathrm{d}t \tag{1.35}$$

Under above equations, we can obtain the heat-to-current efficiency of as-designed heat recuperation setup.

$$\eta = \frac{\int_{t0}^{t1} (V_1 - V_0) I_{\mathrm{dis}}\mathrm{d}t}{(1 - \eta_{\mathrm{HX}}) \sum m_i c_{p,i} \Delta T + T_{\mathrm{H}}\alpha q_{\mathrm{dis}}} \tag{1.36}$$

However, the calculation accuracy of this method still requires attention. As found, the specific heat capacity is typically measured by differential scanning calorimetry (DSC), which cannot rule out the existence of measurement errors. Thus, the performance evaluation for thermoelectrochemical devices should be considered comprehensively with other parameters.

1.3.5 Power Density

Power density, as another key index, is always reported to highlight the breakthrough of as-fabricated devices. However, the units used are relatively different including W, W g^{-1}, and W m^{-2}. This phenomenon not only makes comparisons between different works difficult but also makes it difficult to determine whether there is a real breakthrough. Thus, the normalized power density ($P/(A\cdot\Delta T^2)$) should be provided for reported findings. To date, the value of $P/(A\cdot\Delta T^2)$ is still relatively low (below 3 mW m^{-2} K^{-2}), implying relatively sluggish conversion kinetics from heat to electricity using reported thermoelectrochemical techniques [5, 9, 14, 17–20]. In this case, new materials and devices should be further developed with high energy conversion efficiency and satisfying normalized power density.

References

1 Forman, C., Muritala, I.K., Pardemann, R., and Meyer, B. (2016). *Renewable Sustainable Energy Rev.* 57: 1568.

2 Shi, X.L., Zou, J., and Chen, Z.G. (2020). *Chem. Rev.* 120: 7399.

3 Massetti, M., Jiao, F., Ferguson, A.J. et al. (2021). *Chem. Rev.* 121: 12465.

4 Zhao, D., Wang, H., Khan, Z.U. et al. (2016). *Energy Environ. Sci.* 9: 1450.

5 Han, C.G., Qian, X., Li, Q. et al. (2020). *Science* 368: 1091.

6 Li, Z., Xu, Y., Wu, L. et al. (2022). *Nat. Commun.* 13: 132.

7 Chen, D., Li, Z., Jiang, J. et al. (2020). *J. Power Sources* 465: 228263.

8 Kang, T.J., Fang, S., Kozlov, M.E. et al. (2012). *Adv. Funct. Mater.* 22: 477.

9 Yu, B., Duan, J., Cong, H. et al. (2020). *Science* 370: 342.

10 Meng, T., Xuan, Y., Zhang, X., and Appl, A.C.S. (2021). *Energy Mater.* 4: 6055.

11 Cho, J.Y., Lim, Y.S., Choi, S.-M. et al. (2011). *J. Electron. Mater.* 40: 1024.

12 Parker, W., Jenkins, R., Butler, C., and Abbott, G. (1961). *J. Appl. Phys.* 32: 1679.

13 Buckingham, M.A. and Aldous, L. (2020). *J. Electroanal. Chem.* 872: 114280.

14 Im, H., Kim, T., Song, H. et al. (2016). *Nat. Commun.* 7: 10600.

15 Lee, S.W., Yang, Y., Lee, H.W. et al. (2014). *Nat. Commun.* 5: 3942.

16 Laheäär, A., Przygocki, P., Abbas, Q., and Béguin, F. (2015). *Electrochem. Commun.* 60: 21.

17 Duan, J., Feng, G., Yu, B. et al. (2018). *Nat. Commun.* 9: 5146.

18 Zhang, L., Kim, T., Li, N. et al. (2017). *Adv. Mater.* 29: 1605652.

19 Hu, R., Cola, B.A., Haram, N. et al. (2010). *Nano Lett.* 10: 838.

20 Han, Y., Zhang, J., Hu, R., and Xu, D. (2022). *Sci. Adv.* 8: eabl5318.

2

Conventional Thermoelectric Devices for Low-Grade Heat Harvesting

2.1 Basic Structure and Working Principle of Electronic Thermoelectric Device

The traditional thermoelectric devices are designed in series and in parallel with the "π" structure of P-type materials and N-type materials to ensure the required output power. The main carrier of P-type thermoelectric materials is a hole, and the carrier movement direction is the same as the current direction. The main carrier of N-type thermoelectric materials is electrons, and the carrier movement direction is opposite to the current direction. Therefore, in thermoelectric devices with a fixed hot end and cold end, the series connection of P-type thermoelectric material and N-type thermoelectric material can realize the sequence of current direction and provide a more powerful electrical output. Benefiting from rational thermoelectric designs, significant progress has been made on high-performance and cost-effective thermoelectric materials. Figure 2.1 presents the reported state-of-the-art thermoelectric materials with maximum ZT values (ZT_{max}), cost, and cost-effectiveness (ZT_{max}/cost).

2.1.1 Working Principle of Electronic Thermoelectric Devices

The reason why the traditional thermoelectric material energy can be converted to each other is based on three effects. These three effects are not independent but are linked by Kelvin's relationship. At the same time, they are also the basis for thermoelectric devices to realize the mutual conversion of thermal energy and electric energy.

2.1.1.1 Seebeck Effect

The Seebeck effect is a phenomenon that can convert heat energy into electric energy. The basic principle can be shown in Figure 2.2a. Two different conductor materials A and B are joined to form a circuit. If the temperature of the two joint ends is maintained at T_1 and T_2, respectively. Assuming that there is a certain temperature difference ΔT between the two ends, this will drive the carrier to move

Low-Grade Heat Harvesting: Materials, Devices, and Technologies, First Edition. Xiaogang Zhang.

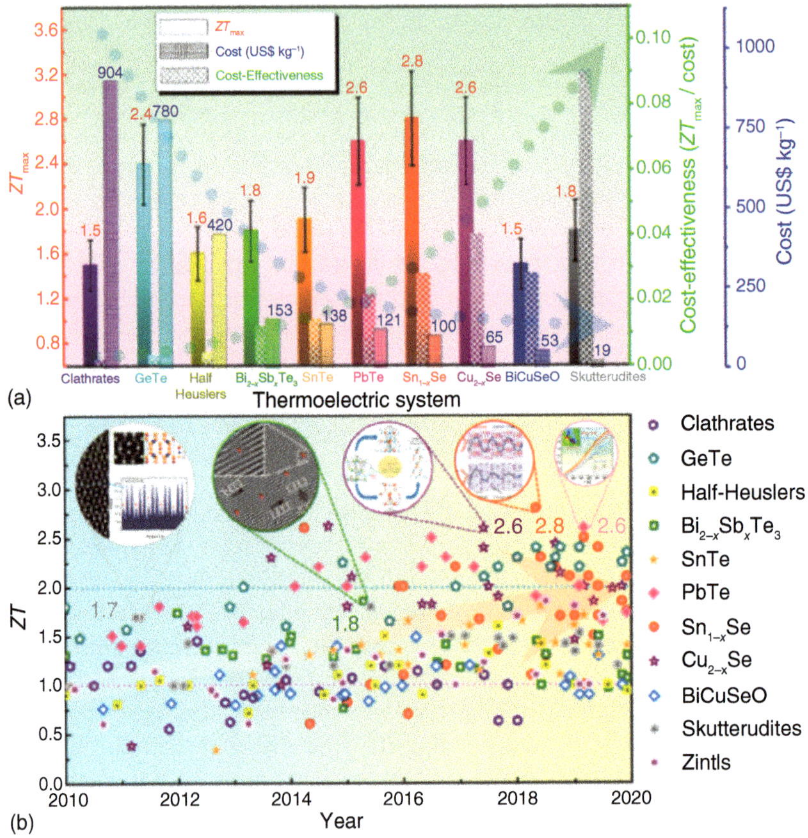

Figure 2.1 *ZT* values for state-of-the-art thermoelectric materials in the last 10 years. (a) Reported maximum *ZT* values (ZT_{max}) with error bars, calculated cost, and calculated cost-effectiveness (ZT_{max}/cost), and (b) timeline for the state-of-the-art thermoelectric materials (bulks). Source: Shi et al. [1]/American Chemical Society.

in a direction, generate current, and form a thermoelectric electromotive force V in the circuit. Therefore, the Seebeck coefficient is defined as:

$$S = \frac{dV}{dT}$$

where S is the Seebeck coefficient, unit: $\mu V\ K^{-1}$, the value can be positive or negative according to the thermal conductivity property of different materials.

2.1.1.2 Peltier Effect

The Peltier effect is the inverse effect of the Seebeck effect. Its principle is shown in Figure 2.2b. If current I is applied to the circuit, there will be heat absorption or heat release at the joint end. Heat absorption or heat release is related to the given current direction, and the rate of heat absorption or heat release is related to the type of material and the current size. The rate of heat absorption or heat release at

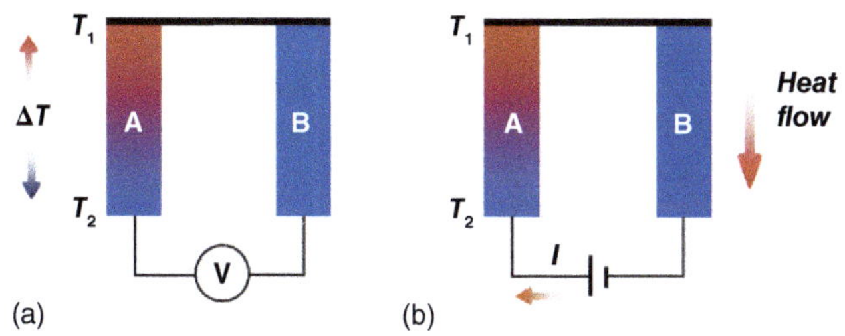

Figure 2.2 Schematic diagram of Seebeck effect and Peltier effect. (a) Schematic diagram of Seebeck effect and (b) Peltier effect.

the joint end is expressed as q, which is proportional to the current I passing through the material in the circuit. The formula is as follows:

$$q = \pi_{AB} I \tag{2.1}$$

where π_{AB} represents the heat absorbed or released by the current at the joint end per unit of time, which is a constant and is the difference between the Peltier coefficients of A and B materials. The unit is usually V or W A^{-1}. The positive value of π_{AB} represents the absorbed heat, and the negative value represents the released heat.

2.1.1.3 Thomson Effect

The Thomson effect is different from the Seebeck effect and the Peltier effect. It does not need two conductors to form a circuit. It is a thermoelectric conversion phenomenon existing in a single uniform conductor. Assume that the current flowing through a uniform conductor is I, and the temperature difference applied in the current direction is $\Delta T = T_1 - T_2$, then the heat absorption (heat release) rate on the conductor is as follows:

$$q = \beta I \Delta T \tag{2.2}$$

In the above formula, β is the Thomson coefficient, in V K^{-1}. Thomson effect is similar to the Peltier effect, but the difference of carrier energy in the loop of the Peltier effect is caused by the difference of carrier potential energy between the two conductors, while the Thomson effect is caused by temperature gradient.

In general, the maximum energy conversion efficiency of thermoelectric devices is determined by the following formula:

$$\eta = \frac{T_\mathrm{h} - T_\mathrm{c}}{T_\mathrm{h}} \frac{\sqrt{1 + Z\overline{\overline{T}}} - 1}{\sqrt{1 + Z\overline{\overline{T}}} + \frac{T_\mathrm{c}}{T_\mathrm{h}}} \tag{2.3}$$

where T_h and T_c are the operating hot and cold end temperatures of the device, respectively, and ZT_ave is the average ZT optimal value within the operating temperature range. The above ZT is a dimensionless quantity to measure the characteristics of thermoelectric materials, $ZT = S^2\sigma T/\kappa$, only related to the intrinsic physical properties of the material, Seebeck coefficient (S), conductivity (σ), and thermal conductivity (κ) relevant. It is not difficult to see that the temperature difference and the ZT_{ave} value of thermoelectric materials during operation are the keys to determine the conversion efficiency of thermoelectric devices. However, the above parameters related to ZT in thermoelectric materials are coupled with each other, which leads to the fact that the ZT value of thermoelectric materials cannot be increased without limitation. In the past 200 years, thermoelectric materials working at different temperature gradients have been extensively studied in order to find the ideal, that is, both large S and σ value and low κ value of thermoelectric material. Among them, the development of energy band engineering to optimize electrical properties and defect engineering to optimize thermal properties has effectively improved the performance of thermoelectric materials. Admittedly, the reported ZT record value has been continuously rising, and the theoretical conversion efficiency of these materials has exceeded 10% (much more than 3–5% of the first-generation thermoelectric generator). However, at the application end, it is still limited to the traditional first-generation thermoelectric generator, for example, the Rover Perseverance, which landed on Mars in 2020, is equipped with a radioisotope generator with thermoelectric devices as the core. However, the thermoelectric materials used in its thermoelectric module are still composed of traditional PbTe and TAGS (short for GeTe-AgSbTe_2 base materials) base materials. For civil use, few corresponding thermoelectric power generation products have been reported.

2.2 Material System of Electronic Thermoelectric Device

Thermoelectric materials can be divided into metal alloys, semiconductor metal oxides, etc. according to their materials. Among them, the existing widely used ones are low-temperature zone, medium-temperature zone, and high-temperature zone. Bi_2Te_3-based thermoelectric materials have large coefficients and low thermal conductivity, and their room temperature electrical merit value is about, which has been recognized as the best thermoelectric materials. In the past few years, Bi_2Te_3 has achieved breakthroughs in combination with new nanotechnology, superlattice technology, thin film technology, etc., making the ZT value reach more than 2. Bi_2Te_3 can also achieve different effects through different doping. The research shows that Ge, Y, and Sm doping has a significant impact on the morphology of the synthesized powder. The possible reason is that rare earth elements change the bond energy after replacing Bi atoms, thus affecting the growth rate in the a, b, and c axes. Ge, Y, and Sm doping has little effect on conductivity and Seebeck coefficient, but it helps to reduce thermal conductivity. Among the three elements, Ge doping can reduce thermal conductivity most effectively. The ZT value of the block after hot pressing can still reach 1.29, which is higher than the value of the commercial area melting block, showing the advantages of combining element doping with nanotechnology.

PbTe is a kind of excellent thermoelectric material. The current preparation method mainly introduces different metals, elements, carbon fibers, etc. to form alloys. Different concentrations of PbI_2 and Na doping can change its thermoelectric properties. With the increase of elements, the power factor shows a downward trend.

SiGe is the best material in this series of alloys. Because of its high thermal conductivity, the single substance Si has low ZT at room temperature. After alloying with Ge, its thermal conductivity is greatly reduced. When $T = 1100$ K, its thermal conductivity reaches the minimum, and its ZT reaches the maximum.

2.2.1 Bi_2Te_3-Based Thermoelectric Material

Bi_2Te_3-based thermoelectric materials were discovered in the 1950s. This material has excellent thermoelectric properties near room temperature and is widely used for refrigeration and power generation near room temperature. It is the only thermoelectric material system that is widely used commercially in thermoelectric materials at present. Bi_2Te_3 compound is a typical narrow-band semiconductor material with a band gap of about 0.15 eV. It is a hexahedral layered structure and has obvious anisotropy. Bi and Te are group V and group VI elements, respectively. The crystal structure of Bi_2Te_3 is a cubic layered structure. It is composed of Bi and Te atomic layers in the order of Te1–Bi–Te_2–Bi–Te1 (as shown in Figure 2.3a), where Bi and Te atoms form a covalent bond, and two adjacent Te atomic layers interact with each other by van der Waals force, so interlayer cleavage is very easy to occur. The energy band structure of Bi_2Te_3 is complex, and the symmetry of the crystal creates six degenerate carrier pockets D1 in the conduction and valence bands, respectively, as shown in Figure 2.3b, which is conducive to improving the density of states and the effective mass of electrons. At the same time, the difference in electronegativity between Bi and Te atoms is small, which helps to obtain higher carrier mobility. For example, its electron mobility along the plane direction can reach 1200 $cm^2 V^{-1} s^{-1}$ at 293 K, so Bi_2Te_3 shows excellent electrical transport performance. In addition, in Bi_2Te_3 material, both Bi and Te atoms have large atomic mass, and the low melting point of the system (about 858 K) makes the system have low lattice thermal conductivity. The combined effect of the two makes it the best thermoelectric material near room temperature.

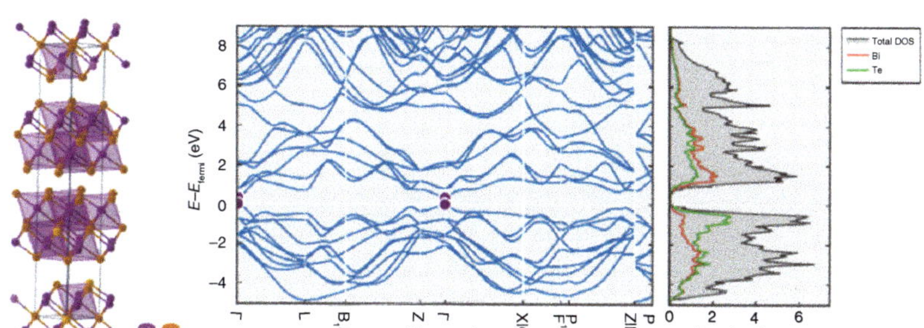

(a) (b)

Figure 2.3 Crystal structure (a) and electron band structure (b) of Bi_2Te_3.

Over the years, there are many ways to improve the ZT value of thermoelectric materials. However, in general, most of the methods are achieved by reducing the lattice thermal conductivity of materials, and there are relatively few methods to enhance the power factor of materials. However, the lattice thermal conductivity of the material has the lowest lower limit, that is, the average free path of phonons cannot be less than the distance between atoms of the material. Therefore, if we want to make a breakthrough in the ZT value of materials, we must find effective ways to improve the power factor of materials.

At present, there are many methods to optimize the power factor of materials by increasing the carrier concentration of materials to increase the conductivity. For example, the donor effect after donor impurity doping promotes more carriers to participate in the transmission by the Fermi level shift guide. However, there are conflicting constraints between the conductivity σ and the Seebeck coefficient. When the carrier concentration of materials is increased in order to obtain greater conductivity, the Seebeck coefficient will decrease, and at the same time, the contribution of carriers to the thermal conductivity will be increased. Therefore, increasing the carrier concentration of the material is not a good idea. As shown in Figure 2.4, the peak value of ZT occurs in the range of carrier concentration of 10^{19} cm^{-3}–10^{20} cm^{-3}. Choosing high conductivity inevitably corresponds to higher thermal conductivity and lower Seebeck coefficient, thus obtaining lower ZT value. On the contrary, it is easier to obtain a high ZT value by optimizing the Seebeck coefficient. The mobility μ of carrier is only related to the conductivity of the material, and the electrical properties of the material can be independently controlled by adjusting its size. This is a good way to optimize the material power factor. Therefore, the main means to optimize the material power factor should be to increase the mobility of the material carrier and Seebeck coefficient.

The unique anisotropic structure of Bi_2Te_3-based materials plays a great role in regulating carrier mobility. The electrical and thermal conductivities in the

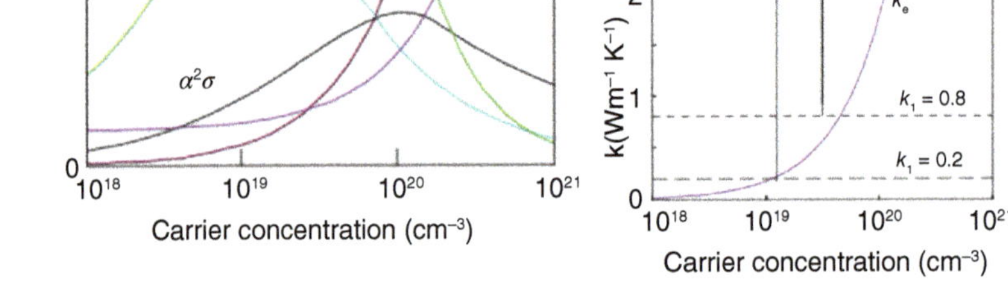

Figure 2.4 The relationship between the hot spot parameter of Bi_2Te_3 and the carrier concentration. Source: Snyder and Toberer [2]/Springer Nature.

vertical C-axis direction are four times and two times that in the parallel C-axis direction, respectively, and the Seebeck coefficient has no obvious anisotropy. The carrier mobility perpendicular to the C-axis is greater than that parallel to the C-axis. Therefore, with the increase of the degree of material orientation, the carrier mobility in the vertical C-axis direction of the material can be maximized. For Bi_2Te_3-based materials, especially N-type Bi_2Te_3-based materials whose thermoelectric properties are more sensitive to orientation degree, it is more effective to improve carrier mobility by adjusting orientation degree. In the Bi_2Te_3-based materials modified with metal elements studied by Wang et al., the doping of Cu elements with different concentrations can increase the mobility of carrier, among which $x = 0.005$, with the lowest concentration, has the best effect (Figure 2.5). Among the many favorable factors, the increase in orientation degree is indispensable. However, in addition to the factors of orientation degree, the "bridging effect" between Cu and adjacent atomic layers alleviates interfacial scattering and also plays a significant role in increasing carrier mobility. Cu atoms are located in the interstitial positions of two "quintets" and are connected to them, which promotes electron transfer in the plane perpendicular to the C-axis. Thus, the fluidity of carriers is enhanced. The final results show that for $Cu_xBi_2(Te_{0.9}Se_{0.1})_3$ material with $x = 0.01$, the power factor of the material reaches a maximum value of 5.3 $mWm^{-1}\ K^{-2}$ when the conductivity and Seebeck coefficient are increased simultaneously.

In 1994, Rowe and Min [4] found that introducing a barrier or potential well into the material, selectively blocking and scattering those carriers with lower energy, will improve the Seebeck coefficient. This is because carriers with different energies

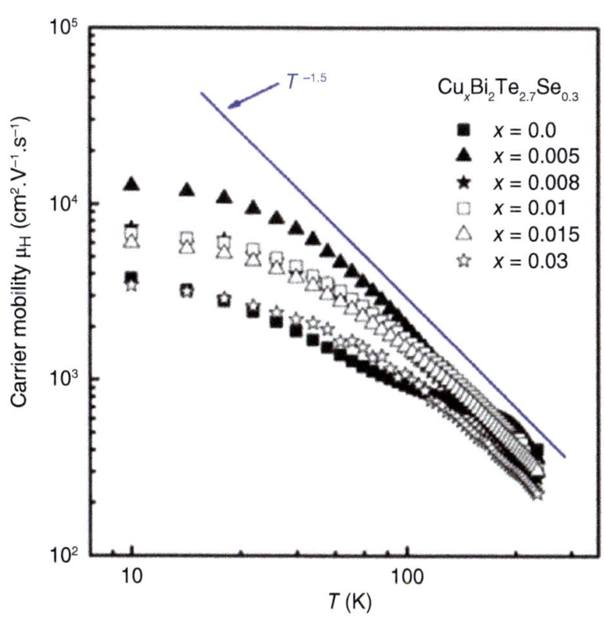

Figure 2.5 Carrier mobility of Cu-doped N-type $Bi_2Te_{2.7}Se_{0.3}$ materials. Source: Wang et al. [3]/IOP Publishing.

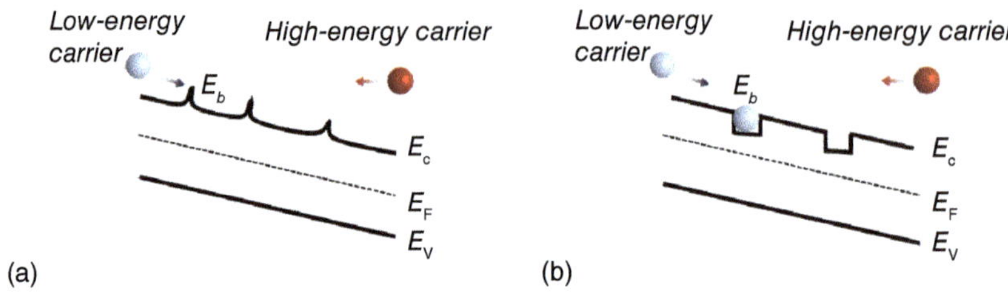

Figure 2.6 Schematic diagram of carrier energy filtration effect. (a) The potential barrier hinders the movement of low-energy charge carriers. (b) The potential well causes scattering of low-energy charge carriers.

have different contributions to the Seebeck coefficient. Carriers with lower energies can not only improve the Seebeck coefficient but also have adverse effects on it. In theory, reducing or even eliminating low-energy carriers will greatly improve the Seebeck coefficient of the material. Figure 2.6 explains the role of potential barrier and potential well in the electron transmission process.

In some theoretical and experimental studies, it has been proved that in the presence of heterogeneous structures and nanocomposites, the Seebeck coefficient is improved due to the introduction of potential well and barrier, and the thermoelectric conversion efficiency is greatly increased. Dou et al. [5] successfully introduced carrier filtering effect into nanocomposites by doping amorphous SiO_2 (α-SiO_2) into $(Bi_2Te_3)_{0.2}(Sb_2Te_3)_{0.8}$ bulk material. The test results show that the Seebeck coefficient and power factor of the material with 0.55% doping concentration are significantly improved, and the *ZT* value of the material is finally 1.12 and 1.27 at about 303 and 363 K, which are increased by about 27% and 20%, respectively. Based on the experience of previous experiments, Hwang et al. [6] used the wet chemistry method to embed metal nanoparticles in the bulk material, and the interface between the nanoparticles and the main material appeared band bending. The barrier was successfully introduced into the P-type $Bi_{0.5}Sb_{1.5}Te_3$ material, which triggered the energy filtering effect of the carrier and obtained the increased Seebeck coefficient. At the same time, nanocomposites containing complex grain boundaries, metal nanoparticles, and other microscopic structures were successfully synthesized, and a wide range of phonon frequency scattering was obtained, which greatly reduced the thermal conductivity of the material. The dual optimization of thermal and electrical properties greatly improved the material performance. The $Bi_{0.5}Sb_{1.5}Te_3$ modified with 0.2% Mn doping concentration reaches the maximum *ZT* value of 1.47 at near room temperature, and good results have been obtained.

2.2.2 PbX (X = S, Se, Te) Compound

PbX (X = S, Se, Te) compound is one of the earliest thermoelectric materials discovered. Seebeck discovered that PbS has the characteristics of a high Seebeck coefficient as early as 1822. However, the systematic study of the physical properties of PbX (X = S, Se, Te) compound is in the period of aerospace development after

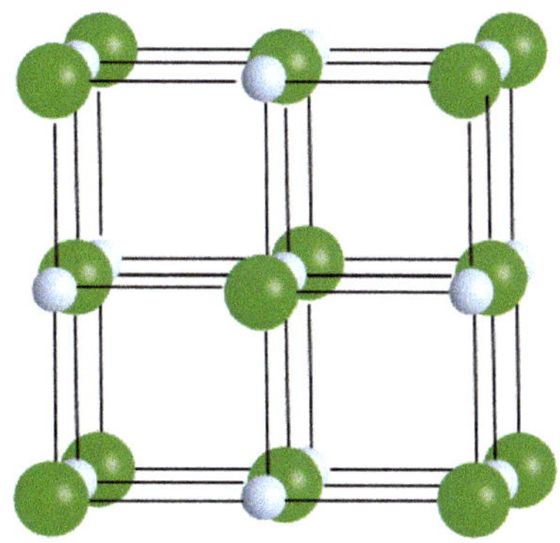

Figure 2.7 Crystal structure of the PbX (X = S, Se, Te) compound.

the Second World War. PbX (X = S, Se, Te) compounds belong to the IV–VI groups. They all have face-centered cubic NaCl crystal structure, Fm3m space group, and the coordination number of each atom is 6, as shown in Figure 2.7. From the appearance, PbX crystal is an opaque and metallic compound crystal with high brittleness and easy cleavage along the (100) crystal plane. The lower the temperature is, the stronger their cleavage trend is, and the cleavage phenomenon basically disappears at high temperature.

PbX (X = S, Se, Te) compound is a kind of polar semiconductor material mixed with ionic bonds and covalent bonds. Early PbX (X = S, Se, Te) compounds have been regarded as typical semiconductor materials dominated by ionic bonds, because their crystal structure is a typical ionic compound structure, the distance between the nearest neighboring atoms is closer to the ionic radius rather than the covalent radius, and their high-frequency dielectric constant and static dielectric constant differ greatly. However, later, in the study of carrier scattering mechanism, it was found that carriers were scattered by optical wave phonons and acoustic wave phonons, and acoustic wave phonon scattering played a major role, which also showed that the bonding of such compounds was mainly dominated by covalent bonds, because in the compounds dominated by ionic bonds, the scattering mechanism of carriers was mainly optical wave phonon scattering. So far, there is still no satisfactory quantitative estimation method to explain the proportion of ionic bonds and covalent bonds in PbX compounds, but the fact that ionic bonds and covalent bonds coexist in PbX compounds is beyond doubt.

People usually obtain N-type or P-type thermoelectric materials by adjusting the stoichiometric ratio of PbX compounds and doping of donors and acceptors. Adjusting the stoichiometric ratio of the matrix material (forming excess Pb or X atoms and Pb or X vacancies) can optimize the carrier concentration of the material, but the solid solubility of excess Pb or Te and Pb or Te vacancies in the PbX matrix material is lower than 0.1%, so the maximum carrier concentration obtained by this method of adjusting the stoichiometric ratio is only 1019 cm^{-3}, which is lower than the optimal carrier concentration required for PbX material as thermoelectric material. By doping at the Pb or Te position, a wider range of regulation can be achieved, such as using Li, Na, K, Rb, and Cs as acceptor doping atoms to achieve P-type doping, in which Na atom doping is the most typical. The solubility of Na atom in PbX materials decreases in the order of S, Se, and Te, and their limit solubility (mass concentration) is 2%, 0.9%, and 0.5%, respectively, while N-type materials can pass Ga, In, La, Sb, Al, Bi, and Cl, Br and I atoms are doped at the Pb position and X position, respectively, where the Zr value of Al-doped PbSe reaches 1.3 at 850 K.

In addition to regulating the electrical properties through N-type or P-type doping, solid solution between PbX (X = S, Se, Te) compounds, namely, the formation of binary (PbS–PbSe, PbS–PbTe, and PbSe–PbTe) or ternary (PbS–PbSe–PbTe) solid solution, is also an effective method to adjust its energy band structure and optimize its thermoelectric properties. PbS–PbSe can form a complete solid solution, while the solid solubility of PbS–PbTe system at low temperature is very limited. The PbSe–PbTe system can form a limited solid solution below 300 °C, and the system can form an infinite solid solution when the temperature is higher than 300 °C until the melting point of PbTe. As shown in Figure 2.8a, in PbTe materials, there is an L valence band with a degeneracy of 4 at the L point of the Brillouin region, and Σ valence band with a degeneracy of 12 at 0.2 eV away from the L valence band. By producing convergence of many valleys at the desired temperature, if properly doped, the L and Σ valence bands of $PbTe_{1-x}Se_x$ can converge, thereby increasing the degeneracy to 16 (Figure 2.8).

In addition to the above-mentioned method of introducing foreign atom doping and forming a solid solution to improve the thermoelectric properties of PbX material system, the introduction of nanostructures in the material system to reduce the thermal conductivity of materials by enhancing phonon scattering is also the main method to optimize the thermoelectric properties of materials. As shown in Figure 2.9, in the PbX material system, the introduced nanostructures are mainly divided into three types: coherent nanostructures, semi-coherent nanostructures, and noncoherent nanostructures. Coherent nanostructure means that there is only a slight mismatch between the nano-materials and the matrix without obvious

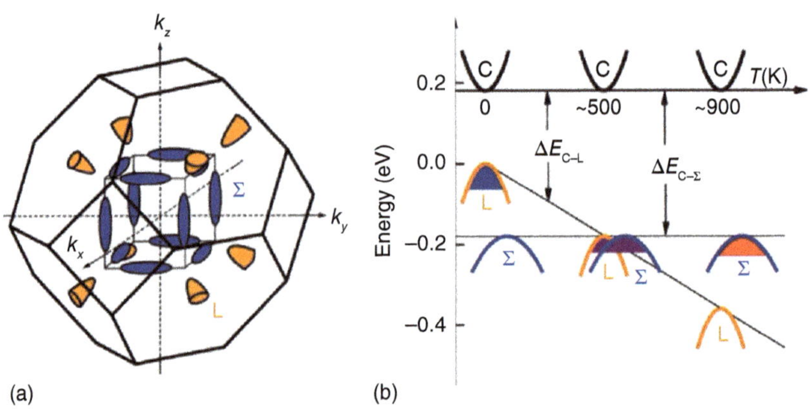

Figure 2.8 Graph of thermoelectric properties of PbTe solid solution compounds as a function of temperature. (a) Brillouin zone showing the low degeneracy hole pockets centred at the L point, and the high degeneracy hole pockets along the Σ line. (b) Relative energy of the valence bands in $PbTe_{0.85}Se_{0.15}$. Source: Pei et al. [7]/Springer Nature.

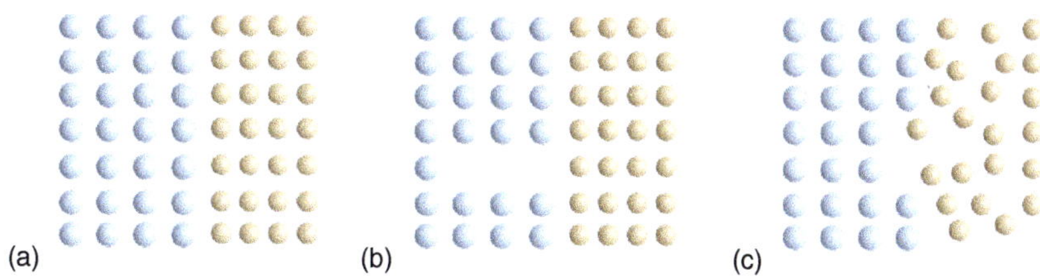

Figure 2.9 (a) Coherent nanostructures, (b) semi-coherent nanostructures, (c) noncoherent nanostructures.

grain boundaries, while semi-coherent structure and noncoherent structure mean that there are obvious grain boundaries between the nano-materials and the matrix. Different nanostructures can reduce the lattice's thermal conductivity by forming different defects and scattering mechanisms, but they also have certain effects on the electrical properties of the material system. Coherent nanostructures act like point defects and scatter short-wave phonons, while semi-coherent nanostructures and noncoherent nanostructures mainly scatter medium-wave and long-wave phonons.

2.2.3 SiGe Alloy

Both Si and Ge belong to group IV elements, and the crystal structure is diamond structure. Although both have high power factors, due to high thermal conductivity, the single substance Si and Ge are not good thermoelectric materials. In the 1960s, it was found that the thermal conductivity of the alloy formed by Si and Ge was significantly reduced, and the reduction of the mobility was far less than that of the thermal conductivity, so the thermoelectric performance of the alloy was much higher than that of Si and Ge. Subsequently, silicon–germanium alloy (SiGe) began to be widely studied and became the most important high-temperature thermoelectric material, which can be used from room temperature to 1300 K.

It can be seen from the Ge–Si crystal structure and electron band structure shown in Figure 2.10 that both can form a continuous solid solution, and their physical

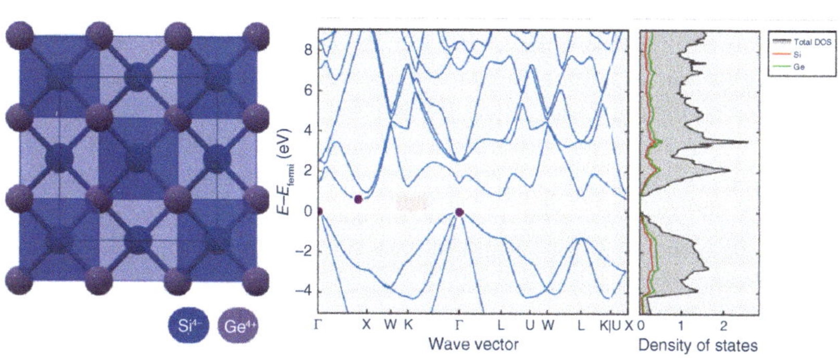

Figure 2.10 Crystal structure and electron band structure of SiGe binary phase diagram.

properties (such as density, lattice constant, melting point, Debye temperature, and band gap width) vary between the corresponding values of the two elements according to the change of alloy composition, so many physical properties of SiGe alloy can be adjusted by changing the composition. At present, the solid solution with high Si content is generally selected as the thermoelectric material of SiGe alloy in practical applications, mainly because the alloy with high Si content has a high melting point and large band gap width, low density, and strong oxidation resistance, which is more suitable for application at high temperature; Moreover, the alloy with higher Si content has lower thermal conductivity, and the solid solubility of many doped atoms in Si is higher, which is conducive to the preparation of heavily doped semiconductors to adjust the thermoelectric properties of materials. Another very important reason is that the price of Ge is expensive, while the cost of Si is lower. At present, $Si_{80}Ge_{20}$ is the most commonly used and mature thermoelectric material component of SiGe alloy.

In order to improve the thermoelectric properties of SiGe alloy, it is necessary to dope it to adjust the carrier concentration. The commonly used donor impurities are group V elements such as P and As, and the commonly used acceptor impurities are group III elements such as B and Ga. The doping concentration corresponding to the optimal thermoelectric properties of SiGe alloy is relatively high, the optimal doping concentration of N-type is 1021 cm^{-3}, and the optimal doping concentration of P-type is 1020 cm^{-3}.

In the 1990s, Dresselhaus and others put forward the view that the thermoelectric properties of materials can be effectively improved through nanotechnology and low-dimensional methods [8]. Since then, improving the thermoelectric properties of SiGe alloy materials through nanoscale structure control has become a new research hotspot. Joshi et al [9] have taken the lead in achieving a breakthrough in thermoelectric properties of SiGe alloy bulk materials through nanostructure. At first, SiGe alloy nanoparticles (particle size of about 20 nm) were prepared by ball milling method, and then sintered and densified by hot pressing method, which reduced the thermal conductivity of the material by about 50% without significantly deteriorating the electrical properties, and finally increased the maximum *ZT* value of P-type SiGe alloy ($Si_{80}Ge_{20}$) from 0.5 to 0.95, and the maximum *ZT* value of N-type SiGe alloy ($Si_{80}Ge_{20}$) from 0.93 to 1.3. The significant improvement of thermoelectric properties is mainly due to the reduction of grain size, which increases the number of grain boundaries, then increases the scattering of phonons by grain boundaries, and reduces the lattice thermal conductivity of materials. Moreover, because carriers and phonons have different mean free paths, that is, nanoparticles do not cause an obvious scattering effect on carriers while scattering phonons strongly, so nanostructured means almost do not affect the electrical properties of materials. In 2012, Bathula et al. [10] adopted a similar mechanical alloying method to prepare N-type silicon–germanium alloy ($Si_{80}Ge_{20}$) with nanostructures. The *ZT* value reached a maximum of 1.5 at 900 °C. In 2015, Bathula et al. again raised the maximum *ZT* value of nanostructured P-type silicon–germanium alloy ($Si_{80}Ge_{20}$) to 1.2 (Figure 2.11).

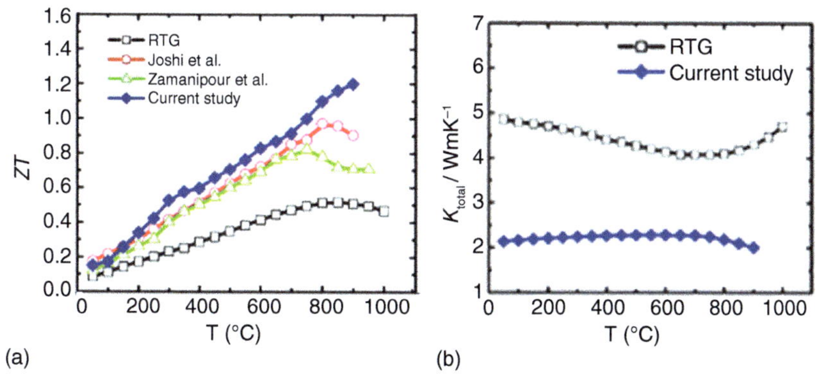

Figure 2.11 Graph of (a) lattice thermoelectric value *ZT* and (b) thermal conductivity of $Si_{80}Ge_{20}$ block thermoelectric material changing with temperature. Source: Bathula et al. [11]/Royal Society of Chemistry.

2.3 Performance and Optimization Method of Electronic Thermoelectric Device

A thermoelectric element usually consists of a metal electrode connecting an N-type thermoelectric material (N-type thermoelectric arm) to a P-type thermoelectric material (P-type thermoelectric arm). Because the output voltage of a single thermoelectric element is very low, it is often necessary to connect many thermoelectric elements in electric series and thermal parallel to form the actual thermoelectric device or component for use. The Π shape plate structure shown in Figure 2.12 is the most typical and most widely used device structure. In the N-shaped structure, the thermoelectric element composed of the N-type and P-type thermoelectric arms of cuboid or cylinder is integrated into two ceramic plates with good electrical insulation and heat conduction in the form of electric series and parallel. The heat flow is transmitted in the direction perpendicular to the ceramic plate, which is suitable for the working environment of flat plate heat source. In the N-shaped plate structure device, the heat flux density in the thermoelectric material is uniform, which is easy to realize one-way heat flow. Through the optimization design of the device structure, it is easy to maximize the thermoelectric performance of the material and achieve high conversion efficiency. However, Π plate structure devices usually have restraint on both hot and cold surfaces when operating, and differences in thermal expansion of different materials caused by longitudinal temperature difference tend to cause large thermal stress, which affects structural reliability.

In addition to traditional Π configurations, several non-π configurations have been raised according to different Π configurations and specific π configurations, the most commonly recognized of which are toroidal and Y-π configurations. In 2002, Weinberg et al. [12] proposed the idea that thermoelectric devices could be designed as ring structures, as shown in Figure 2.13. Type N and type P annular thermoelectric materials (thermoelectric arms) are alternately arranged along the cylindrical heat source coaxial, and a layer of annular isolating material is placed between

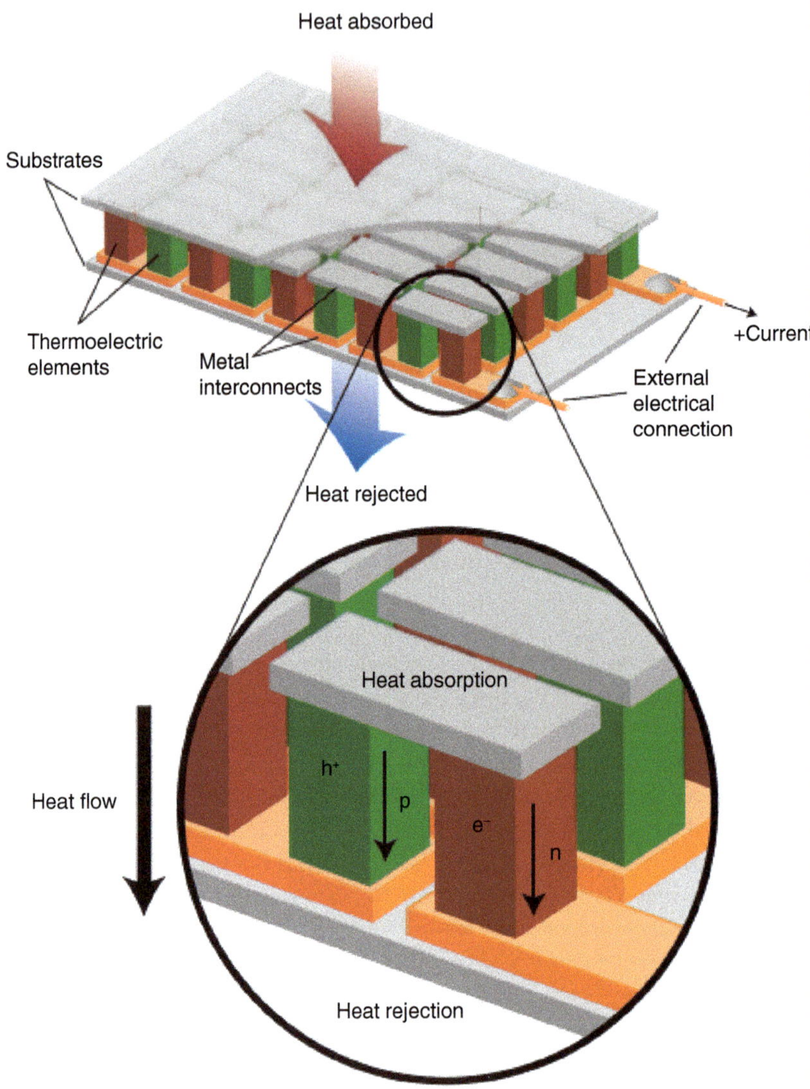

Figure 2.12 Schematic diagram of Π plate structure. Source: Snyder and Toberer [2]/Springer Nature.

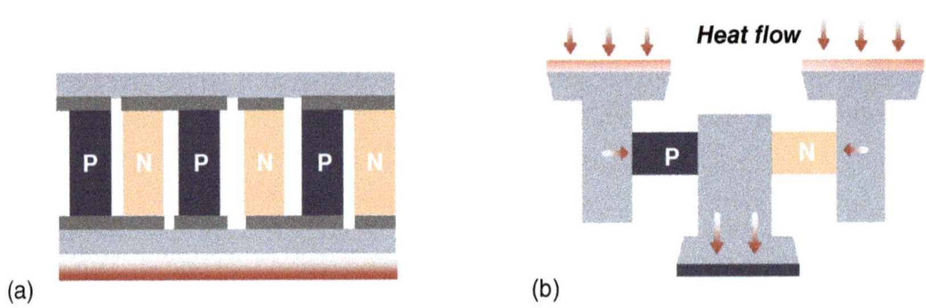

Figure 2.13 Schematic diagram of (a) Π structure and (b) Y structure.

each annular thermoelectric arm to achieve electrical insulation between adjacent thermoelectric arms. In addition, metal ring electrodes are arranged on the outer wall and inner wall of the thermoelectric arm, and the outer and inner metal rings are arranged successively along the axial direction to realize the connection of adjacent thermoelectric arms and the electric series of all p–n thermoelectric elements. This ring-shaped thermoelectric device is suitable for non-plate heat source or fluid medium cold and heat source, especially the heat flow along the radial direction of cylindrical heat source. It is beneficial to overcome the limitation that the traditional N-shaped thermoelectric device can only be applied to the flat plate heat source environment. In addition, compared with the flat structure, it is easier to realize the integrated design and integration of devices and heat collection/heat dissipation components, and easy to realize efficient heat transfer and low-stress structure design. However, the radial transmission characteristics of heat flow and current make it difficult to optimize the design of temperature and electric fields. At the same time, the welding and device integration technology between heterogeneous thermoelectric materials and metal electrodes is more difficult than that of flat devices, and the manufacturing cost is also higher.

Thermoelectric devices generally work in the environment with large temperature difference and cold and hot end constraints, the thermal stress inside the device is inevitable, especially since most thermoelectric materials are brittle materials, stress damage has become one of the outstanding problems in the application of thermoelectric devices engineering [13]. Stress relief or stress-free structure has become the most important goal in device design. Figure 2.13 shows a Y-shaped thermoelectric device first proposed by Bell and Crane in 2006 [14]. In Y-shaped thermoelectric devices, cuboid or columnar N-type and P-type thermoelectric arms are alternately sandwiched between electrode connecting plates in a "sandwich" structure. The electrode connecting plates not only provide a conductive path for adjacent P-type and N-type thermoelectric arms but also constitute a "Y-shaped" heat transmission channel together with P-type and N-type thermoelectric materials as heat collecting and heat transfer components. This Y-shaped structure makes the current flow parallel to the cold and heat sources, i.e. the direction of heat flow is perpendicular to the direction of current. The Y-shaped structure separates the thermoelectric materials from the stress field, and the device installation stress is applied to the high and low-temperature electrodes and insulating parts, but not directly transmitted to the thermoelectric materials or transmitted to the thermoelectric materials after relief. The thermoelectric materials remain relatively independent or not completely constrained, and the thermal expansion deformation is reduced by the cold and hot surface constraint, which is conducive to the stress relief structure design of the thermoelectric materials and structural parts. The transverse series thermal expansion of N-type and P-type thermoelectric arms avoids the stress concentration caused by the difference in expansion coefficients and provides greater flexibility for the height and area design of thermoelectric arms. In addition, the structure also enables each thermoelectric element to be structurally optimized independently. Each thermoelectric arm can have a different height and interface structure, which is conducive to the manufacture of piecewise structured devices. However, the uneven heat flux and current density in the thermoelectric material in Y-shaped devices will affect the maximum performance of the thermoelectric material and reduce the efficiency of the device to some extent. In addition, almost through the device, connecting high and low-temperature electrode plates also cause large heat leakage, which is difficult to improve the conversion efficiency of the device.

2.4 Manufacturing Process of Electronic Thermoelectric Devices

The manufacturing technology of π-structured bismuth telluride thermoelectric devices is well-established and has been widely used commercially. Bismuth telluride devices are mainly manufactured by welding, and the manufacturing process is shown in Figure 2.14. Bismuth telluride devices usually use copper as electrodes and ceramic plates with good thermal conductivity (such as aluminum oxide or aluminum nitride substrate) as electrical insulating plates with hot and cold ends. Usually, the copper sheet is pasted on the ceramic substrate in a certain pattern by direct copper coating and other methods. Then, the P-type and N-type thermoelectric arms are formed by connecting K-type thermoelectric elements in series in a sandwich type (i.e. insulating ceramic sheet/electrode/P-type and N-type thermoelectric material/electrode/ceramic sheet) structure using traditional tin welding technology. In order to improve the wettability of thermoelectric materials and solder, nickel plating is generally done in advance on the welding surface of thermoelectric materials. The fabrication process is simple and low cost, but it also has many disadvantages. For example, due to the melting point of the solder, the device is usually used at a temperature below 200 °C, and the use of ceramic insulation makes the device brittle and has poor mechanical impact resistance.

The application of middle- and high-temperature thermoelectric devices is mainly limited by the high-temperature electrode and its connection surface stability, high-temperature stability, and low interface resistance and interface thermal

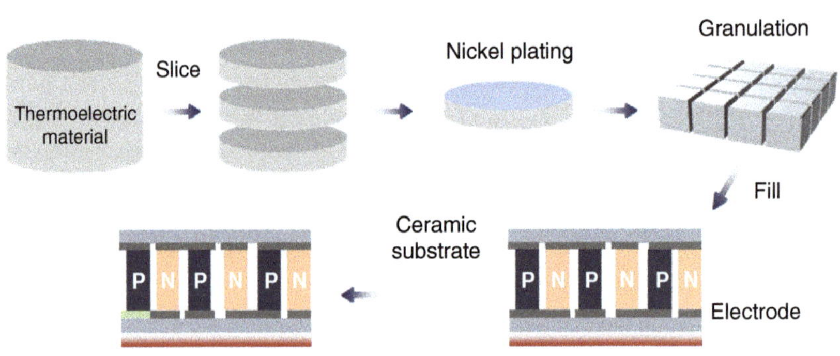

Figure 2.14 Process diagram of Π structure bismuth telluride thermoelectric devices prepared by welding method.

resistance electrode materials and the development of welding technology with thermoelectric materials more and more academic and industrial attention. In recent years, a variety of electrode preparation technologies have been developed. For example, arc spraying, high-temperature diffusion welding, spark plasma sintering (SPS) system one-step combination, and other technologies have been applied to the preparation of thermoelectric devices. Arc spraying technology is characterized by the use of high-temperature resistant frame structure (such as ceramic frame, heat resistant nylon, and arc spraying technology process frame), P-type and N-type thermoelectric materials (thermoelectric arm) placed in the frame, the use of direct current field melting alloy wire, and directly sprayed on the thermoelectric arm after grinding processing technology to complete the electrode manufacturing. By designing the height of the side walls of the frame, it is easy to realize the series/parallel structure flow of the thermoelectric element, as shown in Figure 2.15. Arc spraying technology does not use solder, the use of the device temperature is not limited by solder melting point and high-temperature stability. The bonding state (bonding strength, interfacial resistance, interfacial thermal resistance, etc.) between the alloy electrode and the thermoelectric material mainly depends on the wettability between the molten alloy and the thermoelectric material and the device structure design. In general, to improve the interface bonding state, a transition layer or an intermediate layer is usually prepared in advance at the electrode connection surface of the thermoelectric material. In addition, the thermoelectric arm of arc spraying device is completely fixed and packaged in the space composed of the frame and electrode, forming a complete whole, so the device structure has high strength and good mechanical impact resistance. Arc spraying technology has been successfully applied in the production of bismuth telluride devices.

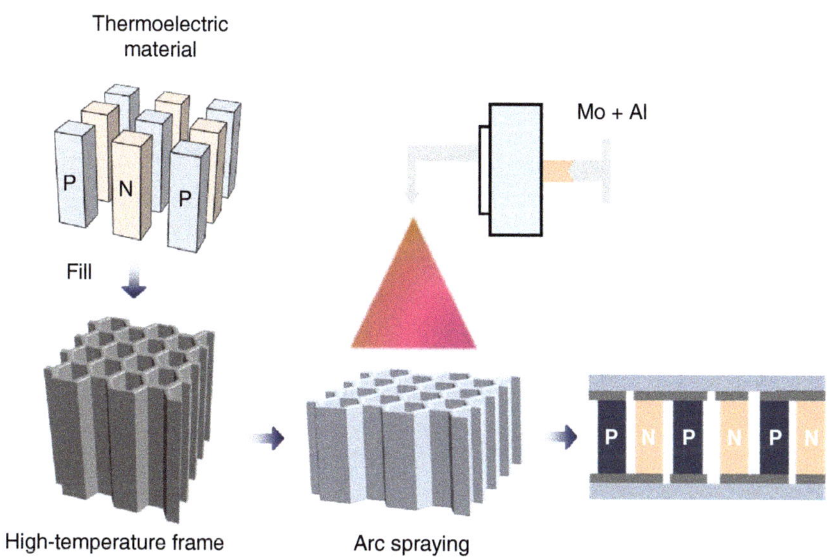

Figure 2.15 Process diagram of Bi_2Te_3-based thermoelectric device prepared by arc spraying technology.

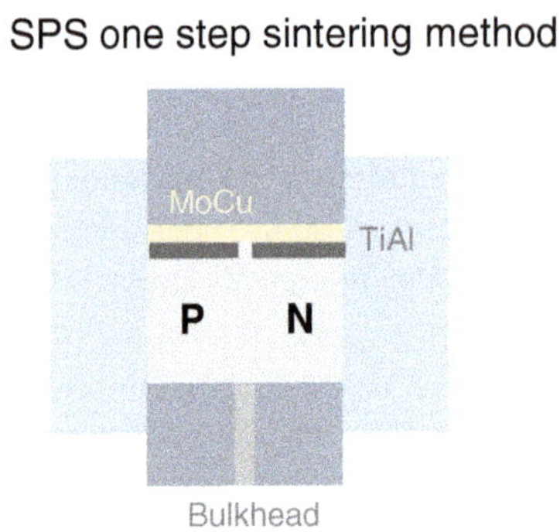

Figure 2.16 Schematic diagram of thermoelectric element SPS one-step preparation.

The above two processes are directly connected and integrated into components by using P-type and N-type thermoelectric arms, and a typical process is to prepare P/N materials for high-temperature electrode interconnection Π, and then a series of n-shaped elements are interconnected to form components (Figure 2.16). The core of this process is the one-step preparation of Π-shaped components, using plasma sintering or hot pressing sintering technology to simultaneously prepare P-type and N-type materials. This integrated process combines hot-pressing one-step manufacturing with high-stability interface bonding technology, so it is easy to obtain high-strength bonding of thermoelectric materials and electrodes, avoid high-temperature impact of subsequent welding of thermoelectric materials, and obtain low interface contact resistance. This process has been applied in the preparation of skutterudite devices and PbTe devices.

2.5 Design, Integration and Application of Electronic Thermoelectric Device

2.5.1 Single-Stage/Multi-Stage Device Structure Design

The energy conversion efficiency of thermoelectric device is not only related to the thermoelectric properties of the material and the temperature difference between the two ends of the material but also depends on the size of the thermoelectric arm. When the P-type and N-type thermoelectric materials that make up the thermoelectric pair have different resistivity and thermal conductivity, different geometric sizes of the two couple arms are required to obtain the best device output performance. In this section, the single-stage cobaltite device and two-stage cobaltite/bismuth telluride power generation device are taken as examples to demonstrate the optimal design method of the device structure and the influence of the main structural parameters and interface parameters on the device performance.

Based on the finite element method, the thermal-electric module in ANSYS Workbench was used to take the Skd-based thermoelectric power generation single stage as an example, and the low-temperature end cold source temperature T_c was taken 25 °C, the heat source temperature at the high-temperature end $T_h = 600\,°C$, according to the actual working environment and known material thermal physical parameters, the heat transfer coefficient between the heat source and the device is 6×10^3 W m^{-2} K^{-1}. The heat transfer coefficient between the cold source and the device is 1.2×10^4 W m^{-2} K^{-1}. The height of the thermoelectric arm and the ratio of the cross-section of the p and *n* thermoelectric arm are optimized and analyzed. The size optimization design can meet the different output performance requirements of the device, while the practical device needs to consider the requirements of the practical application for the device structure design.

2.5.2 Selection of Electrode Material and Electrode Connection Technology

In the design and integration of thermoelectric devices, the connection between electrodes and thermoelectric materials, especially the design and connection of high-temperature electrodes of generator components, is the key technology of device integration. The physical properties of the electrode material itself (thermal conductivity, conductivity, thermal expansion coefficient, etc.) and its matching with thermoelectric materials, as well as the bonding state between the electrode and thermoelectric materials (bonding strength, interface resistance, interface thermal resistance, interface high temperature and chemical stability, etc.) directly affect the efficiency, reliability, and service life of the device. The selection of electrode materials mainly follows the following principles: (i) Electrode materials have high conductivity and thermal conductivity to reduce energy loss; (ii) The expansion coefficient of the electrode material shall be as close as possible to the thermoelectric material connected to it to avoid cracking due to stress concentration caused by the mismatch of the thermal expansion coefficient. Generally, the difference in thermal expansion coefficient between the two should be controlled within 20%; (iii) The interface bonding strength between electrode and thermoelectric material is high, and the contact resistance and contact thermal resistance are low; (iv) There is no serious mutual diffusion or reaction between the electrode and the thermoelectric material within the service temperature range to avoid damage to the thermoelectric material; (v) It has a certain degree of oxidation resistance and high-temperature stability within the use temperature range; and (vi) The combination process of electrode and material is simple. Generally, most metal materials are difficult to meet the above requirements at the same time, especially the high bonding strength, low interface resistance, and high-temperature stability of the interface are difficult to achieve at the same time. Therefore, multi-layer interface layers are usually introduced to meet the above contradictory indicators at the same time.

2.5.3 Thermoelectric Material/Electrode Transition Layer and Interface Structure

For thermoelectric devices that have been in service for a long time, the interface between the high-temperature end electrode and the thermoelectric material is prone to mutual diffusion or chemical reaction of elements, resulting in changes in the composition and structure of the interface, resulting in additional interface resistance and thermal resistance, and resulting in device performance degradation and even device failure. In addition, in the selection principles of most metal electrode materials introduced in Section 2.5.2, the matching of high conductivity, high thermal conductivity, and thermal expansion coefficient with thermoelectric materials should be considered first, and these electrode materials are often difficult to form a good combination with thermoelectric materials directly, that is, it is difficult to meet the six requirements of electrode material selection. At the same time, the effective way to solve the problems of good connection between electrode material and thermoelectric material (high strength, low interface resistance, and thermal resistance) and high-temperature interface stability is to introduce an appropriate transition layer between thermoelectric material and electrode. For the soldering process of low-temperature devices, the main function of this transition layer is to improve the wettability of materials and solder and reduce the interface resistance and thermal resistance. For example, for bismuth telluride devices, nickel plating on the welding surface of bismuth telluride material is the most common and effective method. However, for high-temperature devices, the selection and design of the transition layer are more complex, and three main functions need to be considered: (i) plays the role of "activation layer" in the electrode preparation process, promotes the combination of thermoelectric materials and electrodes, and improves the bonding strength; (ii) forms a stable interface layer after completing the preparation process. The interface layer is used to block the mutual diffusion and chemical reaction between thermoelectric materials and electrodes under high-temperature conditions; (iii) The resistance and thermal resistance of the formed interface barrier layer must also be low enough, that is, it must be a good conductor at high temperature.

2.5.4 Device Application and Service Performance

After more than half a century of development, thermoelectric refrigeration devices have a stable market scale, especially in the field of small-volume refrigeration and local refrigeration technology. At the same time, thermoelectric generator parts have irreplaceable status in military and space exploration and other fields and gradually begin to play their role in the recycling of industrial waste heat. Compared with the traditional refrigeration method, the efficiency of thermoelectric refrigeration is independent of the volume. It is suitable for small-volume refrigeration, and precise temperature control can be achieved by adjusting the voltage or current. The thermoelectric refrigeration device also has the advantages of no compressor, no mechanical moving parts, high reliability, no noise, and no ammonia, freon, and other refrigeration media, which is more in line with the requirements of modern society for green environmental protection. As daily consumer products, portable thermoelectric refrigeration refrigerators, thermoelectric air conditioning seats, small air conditioners, drinking fountains, and wine cabinets are the most typical applications. Thermoelectric refrigeration devices are also widely used in optoelectronics,

electronics, and other fields, such as infrared detectors, photomultiplier tubes, charge coupled device (CCD) cameras, laser diodes, thermostats, dehumidifiers, dew point testers, freezing point meters, etc. In addition, thermoelectric refrigeration is easy to achieve point cooling and precise temperature control, which has promoted the development of relevant medical equipment products, including cryo-scalpel, microtome freezers, microscope freezers, medical low-temperature mattresses, and bioreactors. The cryosurgical scalpel using thermoelectric refrigeration technology does not need to use a compressor or liquid nitrogen. Compared with the traditional cryosurgical scalpel, it is smaller and more flexible. At the same time, the temperature of the scalpel head can reach about −50 °C, which is very suitable for skin and eye surgery. The average life of thermoelectric refrigeration devices has exceeded 3×10^5 hours. The most typical application of thermoelectric generator devices is the isotope thermoelectric battery as a space power source. Under the space vacuum environment, the working life of silicon germanium devices and lead telluride devices exceeds 30 years. In addition to radioisotope thermoelectric cells, the heat generated by nuclear reactor heat source, hydrocarbon fuel combustion heat source, and low-grade industrial waste heat can be directly converted into electric energy using thermoelectric devices. Among them, the application of thermoelectric power generation with low energy density and decentralized low-grade heat energy, such as industrial waste heat, automobile exhaust waste heat, and environmental temperature difference, has been a concern for a long time but has not made substantial progress. In addition to reducing costs, improving device conversion efficiency and service reliability is the biggest challenge. Most thermoelectric power generation applications require that the device can work in an environment with large temperature differences, high temperature, or containing water, oxygen, or even corrosive gas for a long time. For flexible devices, it is also required to be able to work in the folded state for a long time. The key components of the device (thermoelectric materials, electrodes, substrates, etc.) will inevitably cause performance deterioration and functional damage during long-term service. The reasons for the degradation and damage of device performance are diverse and complex, mainly involving: (i) the structural evolution of thermoelectric materials under the service environment (especially high temperature) directly leads to the degradation of thermoelectric properties of materials and the reduction of device conversion efficiency; in addition, the volatilization or oxidation of material components will also directly lead to the degradation and even failure of device performance; (ii) Heterogeneous interface in the device is the main reason for device failure, especially the element diffusion and chemical reaction at the heterogeneous interface under high temperature environment will directly lead to the increase of interface resistance, interface thermal resistance, and even the generation of interface cracks, leading to device performance degradation and failure; and (iii) Large temperature difference, mechanical pressure, mechanical vibration, and other service environments lead to structural damage and even fragmentation of devices. Due to the complexity of device failure mechanism, the establishment of device reliability evaluation method and service life prediction method is still an important challenge.

References

1 Shi, X.L., Zou, J., and Chen, Z.G. (2020). *Chem. Rev.* 120: 7399.
2 Snyder, G.J. and Toberer, E.S. (2008). *Nat. Mater.* 7: 105.
3 Wang, S.Y., Li, H., Lu, R.M. et al. (2013). *Nanotechnology.* 24: 285702.
4 Rowe, D.M. and Min, G. (1994). *AIP Conf.* 316: 339.
5 Dou, Y.C., Qin, X.Y., Li, D. et al. (2013). *J. Appl. Phys.* 114: 044906.
6 Hwang, S., Kim, S.I., Ahn, K. et al. (2013). *J. Electron. Mater.* 42: 1411.
7 Pei, Y.Z., Shi, X.Y., LaLonde, A. et al. (2011). *Nature.* 473: 66.
8 Dresselhaus, M.S., Chen, G., and Tang, M.Y. (2007). *Adv. Mater.* 19: 1043.
9 Joshi, G., Lee, H., Lan, Y. et al. (2008). *Nano Lett.* 8: 4670.
10 Bathula, S., Jayasimhadri, M., and Singh, N. (2012). *Appl. Phys. Lett.* 101: 213902.
11 Bathula, S., Jayasimhadri, M., Gahtori, B. et al. (2015). *Nanoscale.* 7: 12474.
12 Weinberg, F.J., Rowe, D.M., and Min, G. (2002). *J. Phys. D Appl. Phys.* 35: 61.
13 Chen, L., Liu, R., and Shi, X. (2018). *Thermoelectric Materials and Devices*, 1e, 195. Beijing: Science Press.
14 Crane, D.T. and Bell, L.E. (2006). Progress towards maximizing the performance of a thermoelectric power generator. *ICT-25th International Conference on Thermoelectrics*, p. 11. https://doi.org/10.1109/ICT.2006.331259.

3

Polymer-based Thermoelectric Devices for Low-Grade Heat Harvesting

3.1 Introduction

With the rapid development of society, global energy and environmental problems are becoming increasingly severe, posing significant challenges that all humankind must face. Therefore, there is an urgent need for new and efficient energy materials to meet the growing demand. Solar energy is the source of almost all energy on Earth, and photovoltaics (PVs) and concentrated solar power (CSP) represent promising solar energy conversion technologies that are expected to become the most representative clean energy sources in the next 30 years. On the other hand, the current global energy utilization rate is less than 40%, with the majority of energy being lost in the form of heat and difficult to utilize, which invisibly exacerbates the energy crisis. With the rapid development of electronic devices such as flexible wearable devices, there has been further promotion of the transformation of their energy supply methods. The utilization of low-temperature waste heat from industry, environmental heat, and even body heat is gradually giving birth to new energy industries. Therefore, improving the efficient application of thermal energy will be an important development direction in the future energy field [1, 2].

The thermoelectric effect is the mutual conversion of thermal energy and electrical energy. When there is a temperature gradient between two materials of different temperatures, charge carriers in the material will move from the high-temperature region to the low-temperature region, resulting in a charge separation phenomenon and thus generating a potential difference. By utilizing this thermoelectric effect, waste heat can be converted into electricity, enabling energy recovery and utilization, thereby improving energy utilization efficiency [3]. Therefore, the performance of thermoelectric materials has become a key factor affecting the thermoelectric effect. However, in the century following the discovery of the thermoelectric effect, the types of thermoelectric materials were all based on metal materials and the development was slow, with a very limited range of applications. It was not until the emergence of quantum mechanics and semiconductor theory, as well as later the proposal of the phonon–electron crystal theory model, low-dimensional thermoelectric material design strategies, and structure cross-scale regulation methods, that directly promoted the rapid development

Low-Grade Heat Harvesting: Materials, Devices, and Technologies, First Edition. Xiaogang Zhang.

of thermoelectric materials. However, the application range of traditional inorganic thermoelectric materials has been limited due to their high cost, poor stability, and toxicity. In contrast, organic thermoelectric materials have unique advantages: (i) High flexibility and plasticity: Organic thermoelectric materials typically have high flexibility and plasticity since they are composed of molecules or polymers. These materials can be prepared in a flexible form and can be easily bent, folded, and cut, making them suitable for various application fields [4] (ii) Simple preparation process: Compared to inorganic thermoelectric materials, the preparation process of organic thermoelectric materials is simpler because they can use standard chemical synthesis techniques. The synthesis process of these materials typically involves techniques such as liquid-phase or gas-phase reactions, solution impregnation, and electrodeposition, which enables large-scale production and lower cost. (iii) Low cost: Due to the simple preparation process and ease of batch production of organic thermoelectric materials, the cost of these materials is usually lower. In addition, the performance of these materials can also be adjusted by adding different functional groups and polymers to meet the requirements of different application fields. (iv) High efficiency: Organic thermoelectric materials typically have high thermoelectric conversion efficiency due to their low thermal conductivity. This allows them to more effectively utilize thermal energy, resulting in higher electrical energy output [5]. They have advantages in low-temperature and micro-temperature difference power generation and refrigeration applications. Therefore, it has become one of the important energy devices for the new generation of flexible electronic devices.

Research on organic thermoelectric materials started relatively late and is currently in the exploration stage. Early research mainly focused on the study of thermoelectric properties of basic units such as single molecules and polymers. In recent years, with the continuous development of thermoelectric materials, research on organic thermoelectric materials has gradually expanded to include composite materials, nanomaterials, and other areas and has been extensively studied [6, 7]. These studies include preparation methods, thermoelectric properties, charge carrier transport mechanisms, and applications of organic thermoelectric materials in thermoelectric devices.

Although organic thermoelectric materials have broad application prospects, there are still some challenges that need to be addressed [8]. First, the stability of the performance of organic thermoelectric materials needs to be further improved. Due to the diversity of their molecular structure and chemical properties, the stability of the performance of these materials is usually poor and easily affected by environmental factors. Second, the thermoelectric conversion efficiency of organic thermoelectric materials needs to be further improved. Although the thermoelectric conversion efficiency of organic thermoelectric materials is already quite high, their performance still lags behind inorganic thermoelectric materials. Finally, there is a need for more exploration and research in the integration and application of organic thermoelectric materials. Although organic thermoelectric materials have been widely used in some application fields, compared to inorganic thermoelectric materials, their integration and application still require more technical support and research.

Therefore, in this chapter, we will introduce organic thermoelectric materials from three aspects. This includes the basic principle of thermoelectric conversion of organic thermoelectric materials, a focus on the types and design principles of organic thermoelectric materials, and a summary of the construction and functionalization of organic thermoelectric devices.

3.2 Conversion Process and Mechanism

Thermoelectric conversion refers to the mutual conversion between thermal energy and electrical energy, which is achieved by the transfer of heat through thermal transport and the transfer of charge through electronic and phononic transport processes. Compared with inorganic materials, the development of organic thermoelectric materials has a shorter history. Moreover, due to the weak non-covalent interactions such as hydrogen bonds and van der Waals forces between organic material molecules, the charge transport mechanism needs to be further investigated. In addition, the thermoelectric conversion process also involves scattering and electromechanical coupling effects, which further increase the difficulty of studying the thermoelectric conversion mechanism. Therefore, understanding the process and basic principles of thermoelectric conversion in organic thermoelectric materials is not only beneficial for guiding the development of organic thermoelectric materials from a theoretical perspective but also has important implications for the construction and functionalization of devices based on organic thermoelectric materials.

Thermal transport: In solid materials, charge (electrons and holes) and phonons can carry heat during the transport process. Currently, the thermal conductivity of organic thermoelectric materials is generally low ($0.1–1\,\mathrm{W\,m^{-1}\,K^{-1}}$) [9], and it is important to further reduce the thermal conductivity to improve the thermoelectric figure of merit. This requires the simultaneous improvement of the Seebeck coefficient and electrical conductivity while reducing the thermal conductivity. The thermal conductivity of a material can be evaluated by two parameters of electronic thermal conductivity (k_e) and phonon thermal conductivity (k_L), and according to the kinetic theory, thermal conductivity can be expressed as:

$$\kappa = -\frac{Q'}{\nabla T} \tag{3.1}$$

where Q' and ∇T represent the heat flux vector per unit area and the temperature gradient, respectively. As can be seen from the equation that reducing the heat flux can lower the thermal conductivity of a material. It should be noted that in addition to heat flux, factors such as carrier concentration, stacking mode, crystal size, grain boundaries, and impurities in the material can also affect heat transport.

Under the assumption that there is no interaction between particles in the material, the thermal transport of electrons follows an ideal electron gas model, and the relationship between the thermal conductivity (k_e) and the electrical conductivity (σ) can be explained using the Wiedemann–Franz law:

$$\kappa_e = LT\sigma \tag{3.2}$$

where L represents the Lorenz constant ($L \approx 2.4 \times 10^{-8}\,\mathrm{J^2\,K^{-2}\,C^{-2}}$). However, the crystalline regularity of organic materials is often lower than that of inorganic materials, and the electronic thermal conductivity in the process of charge carrier transport is lower than that of inorganic materials. In addition, the introduction of dopants can also affect the electronic thermal conductivity of organic materials. Furthermore, the carriers in organic materials are easily affected by weak intermolecular interactions and the influence of surrounding carriers, forming polarons and making it difficult to comply with the ideal electronic gas model. Therefore, the Wiedemann–Franz law is not applicable to organic thermoelectric materials [10].

In solid materials, the interaction between atoms and their neighboring atoms causes any displacement of one atom in the lattice to affect the transport process of the surrounding atoms, and the sustained vibration of the atoms enables heat transfer. This lattice vibration energy is also known as phonons. For organic thermoelectric materials, Debye's specific heat theory can fully explain the thermal conductivity of phonons:

$$\kappa_{\mathrm{L}} = \frac{1}{3} C_{\mathrm{V}} v l_{\mathrm{p}} \tag{3.3}$$

where C_{V}, v, and l_{p}, respectively, represent the specific heat capacity per unit volume of the material, the sound velocity in the solid and the phonon mean free path [11], and C_{V} is obtained by integrating:

$$C_{\mathrm{V}} = 9NK\left(\frac{T}{\Theta_{\mathrm{D}}}\right)^3 f_{\mathrm{D}}\left(\frac{\Theta_{\mathrm{D}}}{T}\right) \tag{3.4}$$

where Θ_{D} is Debye temperature.

$$\Theta_{\mathrm{D}} = \frac{h v_{\mathrm{D}}}{k_{\mathrm{B}}} \tag{3.5}$$

$$f_{\mathrm{D}}\left(\frac{\Theta_{\mathrm{D}}}{T}\right) = \int_0^{\Theta_{\mathrm{D}}/T} \frac{x^4 \exp(x)}{[\exp(x) - 1]^2} \mathrm{d}x \tag{3.6}$$

Charge transport: Charge transport plays a crucial role in thermoelectric conversion, which is one of the most fundamental physical processes directly affecting thermoelectric parameters. Organic thermoelectric materials are usually composed of small molecules or polymers with π-conjugated structures, and their charge transport properties are closely related to the material's electronic structure. Unlike free charge carriers (electrons and holes) in inorganic materials, charge carriers in organic materials are based on the models of solitons, polarons, and bipolarons, thus explaining the material's electrical behavior. The thermoelectric properties of materials exhibit a certain dependence on temperature, this is due to the weak intermolecular forces between organic molecules, which cause thermal driving to induce molecular vibrations and dynamic disorder that affects the mobility of charge carriers. Charge carriers in organic materials are usually confined within the material molecules and cannot migrate freely, resulting in low intrinsic carrier concentration. At the same time, the doping strategy is a basic method and increasingly attracting attention to enhance the mobility of charge carriers, increase the carrier concentration, and optimize the thermoelectric performance of organic materials [8, 11].

For doped organic thermoelectric materials, the carrier hopping behavior typically follows the variable range hopping (VRH) model [12], and the relationship between the electrical conductivity and thermoelectric performance parameters with temperature is defined as follows:

$$\sigma_{\mathrm{VRH}} = \sigma_0 \exp\left(-\frac{T_0}{T}\right)^{1/(\gamma+1)} \tag{3.7}$$

$$S_{\mathrm{VRH}} = \frac{k_{\mathrm{B}}^2}{2\mathrm{e}}(T_0 T)^{1/2} \frac{\mathrm{dln(g)}}{\mathrm{dE}}\bigg|_{E=E_{\mathrm{F}}} \tag{3.8}$$

where σ_0, T_0, κ, and γ represent the pre-conductivity factor, characteristic temperature, thermal conductivity, and transition index (2D and 3D systems are 0.5 and 0.25), respectively. As the temperature increases, the conductivity and Seebeck coefficient generally show an upward trend. Therefore, the thermoelectric performance can be improved within a certain range by raising the temperature:

$$ZT = \frac{S^2 \sigma T}{k} \tag{3.9}$$

In the traditional understanding, organic materials are good insulators, lacking enough freely mobile charges, and the doping strategy adopted can effectively control the concentration of carriers. The purpose of doping is to introduce additional charge carriers by inducing electron transfer between material host and dopant to improve the conductivity of the material. In traditional inorganic semiconductor materials, doping is to achieve the purpose of adjusting the energy level structure of the compound by replacing some atoms in the material. However, since the carrier of the organic material system adopts the hopping transport model, the doping method and mechanism of the organic material are different from those of the inorganic semiconductor material. When the dopant and the host organic material are mixed, the charge transfer between molecules (atoms), the induction of carriers, the enhancement of the interaction between the charges, and the polaron coupling process of the transition state led to the localized state and energy level of the material changes can be made to realize the adjustment of the hot spot performance of organic thermoelectric materials.

Generally speaking, for a dopant with a sufficiently high electron affinity (EA), electrons will transit from the highest occupied molecular orbital (HOMO) energy level of the organic material host to the lowest unoccupied molecular orbital (LUMO) energy level of the dopant to form organic material cations and dopant anions, this kind of doping is called p-type doping [13]. In contrast, when the dopant has a low ionization potential, it induces n-type doping, and electrons transition from the dopant's HOMO level to the LUMO level of the organic material (Figure 3.1) [14]. Therefore, the energy level alignment between the dopant and organic material molecules is crucial. Typically, p-type dopants with high EA (5.2–6.0 eV) and n-type dopants with low ionization potential (2.8–3.6 eV) are more effective for efficient doping of organic thermoelectric materials [15], and the widely

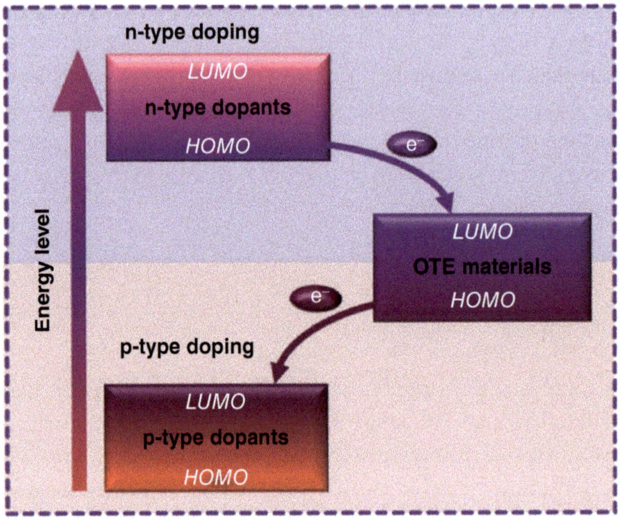

Figure 3.1 The electronic transfer diagrams for p-type doping and n-type doping. Source: Zhou et al. [14]/John Wiley & Sons.

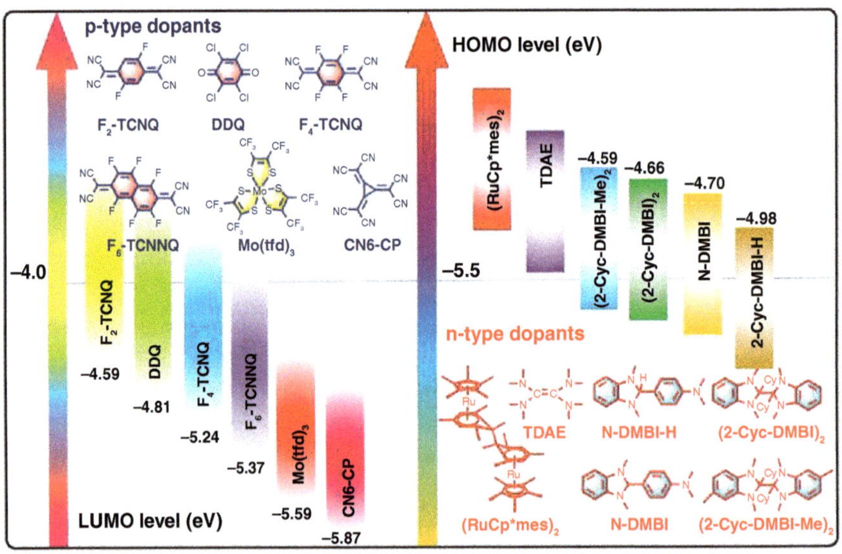

Figure 3.2 Molecular structure and energy levels of common dopants. Source: Zhou et al. [14]/John Wiley & Sons.

used dopants are shown in Figure 3.2 [14]. In the past few decades, p-type dopants and materials have made rapid progress, for example, 2,3,5,6-tetrafluoro-7,7,8,8-tetrafluorotetracyano-quinodimethane (F_4-TCNQ) with an EA (5.24 eV) is the most widely used p-type dopant, and hexacyano-trimethylene-cyclopropane (CN_6-CP) is a small molecule p-type dopant with the highest reported EA (5.87 eV) to date, and is considered a highly promising dopant material [16]. Chabinyc et al. studied poly(2,5-bis(3-tetradecylthiophen-2-yl)thieno[3,2-b]thiophene) (PBTTT-C_{14})

films doped with F_4-TCNQ, it exhibits excellent thermoelectric performance with a power factor as high as $120\,\mu W\,m^{-1}\,K^{-2}$ [17]. Compared to p-type dopants, n-type dopants are faced with the issues of poor air stability and solubility and therefore are used less frequently. In addition, the EA of n-type organic thermoelectric materials is not high, and achieving high fractional doping levels presents a great challenge [18, 19]. Therefore, exploring rational n-type dopants is crucial for improving the thermoelectric performance of n-type compounds.

Doping can adjust the carrier concentration of organic thermoelectric materials, improve their conductivity, and optimize their thermoelectric performance. At the same time, there are still a series of issues regarding the materials, methods, and mechanisms involved in the doping process that need further research. So far, research on organic thermoelectric materials has mainly focused on designing new organic materials (polymers, small molecules, and composite materials) from a molecular level and optimizing existing materials to improve their thermoelectric performance.

3.3 Current Material Types and Design Principles

Organic thermoelectric materials usually consist of conductive polymers, since common polymers are electrically insulating and cannot be used as thermoelectric materials. In the last 20 years, the properties of organic thermoelectric materials have been greatly improved by the discovery of conductive polymers. Conductive polymers exhibit better thermoelectric properties than insulating polymers due to the nature of the chemical bonds in their skeleton. In an insulating polymer, the carbon atoms in the main chain are sp^3 hybridized, and the four electrons in the hybridized orbital are covalently bonded to the other four atoms to form four s bonds. Unlike insulating polymers, conductive polymers have a skeleton composed of sp^2 hybrid carbon atoms. The three electrons in the hybrid orbital participate in forming the three σ bonds that form the backbone of the conducting polymer chain, while the other electron in the unhybrid 2p orbital joins together to form the π bond. Conjugated π bonds are formed when all carbon atoms in the main chain are sp^2 hybridized, where 2p electrons are delocalized and can travel freely between atoms. It is due to the presence of conjugated π bonds that conductive polymers have essentially high conductivity, making them suitable for thermoelectric materials.

According to the type of charge carriers, organic thermoelectric materials can be divided into p-type and n-type, the main charge carriers of which are holes and electrons, respectively. At present, the research on p-type organic thermoelectric materials is more extensive, and the overall performance of the materials is higher than that of n-type organic thermoelectric materials. n-Type organic thermoelectric materials face the problem of poor stability in the air, which brings great difficulty and challenge for related research and practical application.

The trends of high-performance polymer-based thermoelectric materials can be divided into three categories: poly(3,4-vinyl dioxyethiophene) (PEDOT) derivatives, carbon nanotubes (CNTs)/conductive polymer composites, and inorganic semiconductor nanomaterials/polymer composites. Power factor (PF) is often used to estimate the thermoelectric properties of organic materials because of their low thermal conductivity and insensitivity to increases in conductivity.

3.3.1 p-type Organic TE Materials

Conjugated polymer: Polymer has the characteristics of easy preparation, low cost, solution processable, flexible and light weight, which is of great significance to the final application of functional materials. Therefore, it can be predicted that polymer materials will remain the research hotspot of P-type organic thermoelectric materials in the future for a period of time. Polymer-based thermoelectric materials can be roughly divided into polythiophene (PTh), polyaniline (PANI), polypyrrole, polycarbazole, and other conjugated polymer materials.

PTh and its derivatives have attracted great attention because of their excellent photoelectric performance, good stability, solution-processable characteristics, and the structure and property adjustment space brought by different substitution groups on thiophene unit 3 and 4-carbon and have been widely studied in light-emitting diodes, field effect tubes, hydrogen storage, biosensors, and other fields. In addition to conventional factors such as morphology, doping level, and crystallinity, the side chain (type and length) of PTh materials can have an important impact on their transport and processing properties. P3HT and PEDOT are two typical PTh material systems obtained by side chain substitution on thiophene units. Among them, P3HT has attracted wide attention due to its excellent performance in the organic solar cell neighborhood, while PEDOT has become one of the highest-performance organic thermoelectric material systems. Optimization methods for PEDOT derivatives will be introduced in Section 3.3.3.

Although the synthesis and transport performance of PANI has a longer history than that of polymeric thermoelectric materials, compared with PEDOT, the thermoelectric performance and attention of polyaniline are in a relatively weak position. The main reason may be attributed to the fact that polyaniline cannot achieve ultra-high conductivity such as that achieved in the PEDOT system (the conductivity of the PEDOT system can exceed $5000\,S\,cm^{-1}$, while the conductivity of polyaniline system is usually lower than $1000\,S\,cm^{-1}$). Low conductivity not only directly affects the power factor of polyaniline but also restricts the regulation of the Seebeck coefficient of the system by various means (such as the common means in the PEDOT system). Seebeck coefficient optimization usually reduces the conductivity of the system, but the low conductivity of polyaniline makes the effect of this optimization strategy greatly weakened. However, after years of research and optimization, the thermoelectric performance of polyaniline system has been greatly improved, and the power factor can exceed $30\,\mu W\,(m\,K^2)^{-1}$. The arrangement of molecular chains plays a key role in the electrical transport performance of polymer. Thanks to the increase of carrier mobility brought by the orderly arrangement of polyaniline molecular chains in micro and nanostructures, the conductivity and Seebeck coefficient of CSA-doped polyaniline (PANI-CSA) are improved, and the power factor is about 20 times that of amorphous polyaniline.

In addition, the nanostructure, the type, and concentration of dopant all affect the thermoelectric properties of polyaniline system. For example, the conductivity and Seebeck coefficient of polyaniline nanotubes are higher than those of amorphous polyaniline with the same doping level, and the thermal conductivity is lower. The increase in electrical conductivity and Seebeck coefficient is attributed to the increase in mobility caused by the ordered arrangement of polyaniline molecules in the nanotube structure, while the decrease in thermal conductivity is caused by the phonon scattering caused by the nanostructure.

As a common organic semiconductor polymer material, poly (HCHL) attracted people's attention as early as the 1980s. In recent years, with more attention paid to the thermoelectric properties of organic materials, some studies began to pay attention to the thermoelectric properties of poly (Pirochl). At present, the thermoelectric properties of polypyrrole are not as good as those of hot organic thermoelectric materials such as PEDOT. At present, the pure (doped) polypyrrole system has not achieved more competitive thermoelectric performance, mainly because its conductivity is lower than that of polyphene and polyaniline systems on the whole, and the Seebeck coefficient is also at a low level.

Metal–organic complex: The thermoelectric properties of metal–organic coordination polymer poly(M-ett) were first reported in 2012. The type of coordination metal can affect the transport mechanism of poly(M-ETT): poly(Ni-ett) exhibits record n-type organic thermoelectric properties when M is Ni. When M is Cu, the material exhibits p-type organic thermoelectric properties. Furthermore, the two sulfur atoms in ett are replaced by selenium atoms using the atom substitution strategy to obtain the new ligand of dsedt. Both Cu(I)-dsedt and Cu(I)-dsedt showed p-type transport characteristics. Cu(I)-dsedt has the highest power factor and thermoelectric value of 16.4 μW $(m\,K^2)^{-1}$ and 0.025(380 K), respectively. Cu(I)-dsedt has the highest power factor and thermoelectric value of 13.6 μW $(m\,K^2)^{-1}$ and 0.013 (360 K), respectively [20].

Phthlocyanine (Pc) is a kind of typical metal–organic small molecule complex semiconductor materials. The Seebeck coefficient of MPc (M is metal) can reach 10^3 μV K^{-1}, and the Seebeck coefficient of MPc in α phase and β phase will show a significant difference. Meanwhile, in metal phthalocyanine complexes, thc type of metal and the substituent group on the phthalocyanine molecule will affect the size and even the sign of the thermoelectric potential of the material. Since the intrinsic conductivity of phthalocyanine materials is low, it is necessary to improve the conductivity of materials by means of chemical doping. For molecular conductors containing phthalocyanine units, higher power factors can be achieved due to their higher conductivity.

Organic small molecules: Currently, research on p-type organic thermoelectric materials focuses more on polymer and composite and hybrid material systems. On the one hand, these systems currently show better thermoelectric properties, thus stimulating more research interest and enthusiasm. On the other hand, polymer materials have more advantages in large-scale preparation, mechanical properties, and property stability. However, the study of the thermoelectric effect of small molecular materials is also of great significance. First of all, the overall

Table 3.1 Thermoelectric properties of some p-type polymer materials.

Polymer	Dopant	σ (S cm^{-1})	S (μV K^{-1})	PF (μW m^{-}K^{-2})	κ (W m^{-1} K^{-1})	ZT	References
PEDOT	Tos		—	324	0.37	0.25 (300 K)	[21]
PEDOT	PSS		—	469	—	0.42 (300 K)	[22]
PBTTT	F_4-TCNQ	670	42	120	—	—	[17]
PTh	—	565	42	100	0.33	0.1 (300 K)	[23]
P3HT	$Fe(TFSI)_3$	—	—	62.4	0.23	0.1 (365 K)	[24]
P3HT	Mo(tfd-$COCF_3)_3$	12.7	112	16	—	—	[25]
PCDTBT	$FeCl_3$	160	34	19	—	—	[26]
PDPP3T	$FeCl_3$	52	217	247			[27]
PDPPSe-12	$FeCl_3$	—	—	364		0.25 (328 K)	[28]
poly[Cu(I)-ett]		314.4	61.3	118.2	0.79	0.06 (400 K)	[29]

performance level of organic thermoelectric materials is still lower than that of inorganic materials system, so extensive research on different materials is beneficial to explore new high-performance organic thermoelectric materials system. Second, small molecule materials can often adapt to more processing methods, such as solution processing and vapor deposition, which can enrich the design strategy and construction means of organic thermoelectric devices. More importantly, small molecule systems often have more clear chemical structures and more simple and ordered stacking methods, which can often achieve accurate measurement and regulation of physical properties.

Table 3.1 lists the properties of some p-type polymer thermoelectric materials, including conductivity σ, Seebeck coefficient S, power factor PF, thermal conductivity κ, and thermoelectric superiority ZT.

3.3.2 n-Type Organic TE Materials

High efficient thermoelectric conversion is inseparable from the synergistic effect of high-performance p-type and n-type materials, so the development of high-performance n-type materials is very important for the construction of efficient organic thermoelectric devices. In p-type organic thermoelectric materials, the thermoelectric optimum value of PEDOT, diketopyrrolopyrrole (DPP), and other systems has exceeded 0.2. p-type composite material system based on PEDOT has also obtained excellent thermoelectric properties, while related research on n-type organic thermoelectric materials is still faced with multiple challenges such as performance improvement, stability, and machinability. Apart from the performance improvement problems common to organic thermoelectric materials, for n-type organic TE materials, the realization of stable and high performance in the air environment has become one of the most important problems to be solved in the development of efficient n-type organic TE materials. Based on current research, the stability of n-type organic TE materials can be improved by several strategies, such as rational molecular structure design, suitable dopant selection, and appropriate packaging of materials. In the current study of n-type organic TE effect, metal–organic coordination polymer poly(Ni-ett) represents the highest level of n-type organic materials at present, with a thermoelectric value up to 0.3, but it is still very important to explore a wide range of materials and systems. The different characteristics of different materials such as conjugate polymers, metal–organic complexes, and organic conductors in the study of thermoelectric effect are of great significance to the development of n-type organic thermoelectric materials. For n-type polymer materials, although the stability is still a problem to be solved, the advantages of large area preparation and processing of polymer materials, light weight, and low cost will bring great advantages for later practical applications. In addition, the designable space in the molecular structure (skeleton, side chain) of the polymer provides an effective means to optimize the thermoelectric properties. For small organic molecules, although their n-type thermoelectric properties are still lower than the highest level, such materials usually have a more specific molecular structure and accumulation mode, which is more conducive to exploring the structure–activity relationship of the thermoelectric properties of organic materials, providing important basis and guidance for the improvement of the theory of organic thermoelectric effect and the improvement of material properties. Similar to p-type polymer thermoelectric materials, n-type thermoelectric materials can be divided into conjugated polymers, metal–organic complexes, and organic small molecules.

Conjugated polymer: For conjugate polymer materials, the optimization of n-type thermoelectric properties and the enhancement of air stability need to be studied from many aspects, such as molecular design, doping system, and processing means. In the aspect of molecular design, the structure of molecular skeleton will directly affect the assembly process of molecules, resulting in differences in packing mode and material morphology, and thus affect the intramolecular and intermolecular transport of charge carriers. The twisted molecular skeleton is considered to be unfavorable to the efficient transport of carrier within the molecule, so the assembly of the rigid conjugate skeleton has become a hot idea to optimize the thermoelectric properties of n-type copolymers recently. In addition, the optimal design of the molecular skeleton can effectively improve doping efficiency and doping stability. In addition to the skeleton structure of polymer, the side chain structure will also affect the thermoelectric properties of polymer. Reasonable side chain design can enhance the miscibility between polymer and dopant and improve doping efficiency. The choice of dopant and doping mode is also important for the realization of high thermoelectric properties: the dopant needs to have good miscibility and matching energy levels with the subject material, while the doping process needs to reduce the interference to the ordered arrangement of the subject molecules, and the degree of doping needs to be precisely regulated to achieve the best performance. In addition, the thermoelectric stability of n-type polymer materials is an important index to determine the practical value. Various packaging measures can achieve the effect of improving stability. But the structural optimization and rational design of the material itself may be the fundamental means to solve the stability problem.

Metal–organic complex: Metal–organic coordination polymers have been mentioned in the previous section. Poly (M-ett) is a coordination polymer based on vinyl tetramercaptan ligand and metal (M is metal, ett is tetravinyl ligand). When M is Ni, the Seebeck coefficient of poly (Ni-ett) is negative, showing n-type thermoelectric properties. The electrical conductivity and Seebeck coefficient of the powder compactor at room temperature are $44\,S\,cm^{-1}$ and $-122\,\mu V\,K^{-1}$, respectively, and the power factor reaches $66\,\mu W\,m\,K^{-2}$. The thermoelectric superiority value of poly (Ni-ett) at 440 K reaches 0.2, which makes it become a potential n-type organic thermoelectric material. However, some problems of this system need to be solved in order to promote its further development in the field of organic thermoelectric materials: First, the solubility of such materials is poor, limiting the corresponding research and processing means; Second, the powders prepared by liquid phase "one-pot reaction" method show obvious amorphous characteristics, and too disordered molecular arrangement may prevent the materials from obtaining higher thermoelectric properties. Finally, the reaction conditions, reaction mechanism, material structure, and other issues in the preparation process of such materials need to be explored more systematically and deeply, so as to help researchers better understand and optimize the thermoelectric properties of the system.

Organic small molecules: Early studies on the thermoelectric properties of n-type organic small molecules mainly focused on the charge transfer complex represented by TTF-TCNQ. This kind of material usually has a high conductivity, with the inherent advantages of thermoelectric properties research and optimization. At the same time, their clear crystal structure is conducive to the study of structure–activity relationship of organic thermoelectric effect. For organic small molecules materials, the thermoelectric properties are closely related to the molecular packing mode: single-crystal materials have ordered and compact molecular packing and the resulting excellent transport properties, but the size and mechanical strength of organic single-crystal materials make it difficult to characterize the properties and practical applications. Powders or films are more conducive to the characterization of properties and are more suitable for current application solutions, but their properties will be damaged by the disorder of molecular accumulation.

At present, the performance of n-type organic thermoelectric materials has advanced with that of p-type organic materials. The thermoelectric value of poly (Ni-ett), A-DCV-DPPTT, and other materials has reached or approached 0.3. The thermoelectric properties of some n-type organic materials are shown in Table 3.2. However, there are still many challenges in related research. In addition

Table 3.2 The thermoelectric properties of some n-type organic materials.

Polymer	Dopant	σ (S cm^{-1})	S (μV K^{-1})	PF (μW m^{-1} K^{-2})	κ (W m^{-1} K^{-1})	ZT	References
P(NDI2OD-T2)	N-DMBI	8×10^{-3}	−850	0.6	—	—	[30]
PNDTI-BTT-PD	N-DMBI	5	−169	14.2	—	—	[31]
P(NDI2OD-Tz2)	TDAE	-	—	1.5	—	—	[18]
BBL	TDAE	—	—	0.43	—	—	[32]
FBDPPV	N-DMBI	14	−210	28	—	—	[33]
PDPE	N-DMBI	1.3	−235	4.65	—	—	[34]
LPPV-1	N-DMBI	1.1	—	1.96	—	—	[35]
Poly (Ni-ett)	—	200–400	−90 to −140	453	0.4–0.5	0.3 (300 K)	[36]
(TTF) [Ni(dmit)$_2$]$_2$	—	300	−31	29		0.25 (328 K)	[37]
TTF-TCNQ	—	500	−28	39.2			[38]
A-DCV-DPPTT	N-DMBI	5.3	−665	236	0.34	0.26 (373 K)	[39]
TDPPQ	Bi	3.3	−585	113	0.79	0.06 (400 K)	[40]
2DOQT	N-DMBI	14	−183	41.6	0.42	0.03 (300 K)	[41]

to improving the thermoelectric performance of materials, how to improve the stability in the air and obtain the ideal service life has become one of the core issues in this field. To solve this problem, we can not only improve the stability of doping by starting with the design and selection of the main molecule and dopant molecule but also develop simple and effective encapsulation technology to isolate the material from the air environment and improve the stability of the material. In terms of performance optimization, the research strategy of n-type organic thermoelectric materials has something in common with that of p-type materials. First, based on the microscopic mechanism of thermoelectric conversion process, the thermoelectric properties of materials should be improved through targeted molecular design, assembly regulation, doping regulation, and other means. Second, it is necessary to actively develop accurate and efficient performance characterization means, accelerate the exploration of thermoelectric performance of existing high-performance n-type organic semiconductor materials, and accelerate the exploration of new material systems. Finally, it is necessary to conduct targeted optimization of the design and construction of organic thermoelectric devices, so as to improve device performance on the basis of material performance improvement and fully explore the application potential of organic thermoelectric devices.

3.3.3 PEDOT Derivatives

PEDOT is a typical p-type thermoelectric material. PEDOT and its derivatives have high electrical conductivity and good stability among a large number of conductive polymers, showing excellent performance. Although PEDOT is an insoluble polymer, the solubility limit can be reduced by using the water-soluble polyelectrolyte polystyrene sulfonic (PSS) acid as a charge balancer during polymerization. In addition, PEDOT: PSS has excellent film-forming properties, high conductivity, high visible light transmittance, and excellent stability. The colloidal dispersions of PEDOT:PSS nanoparticles in water has been commercialized. Due to the above advantages, the thermoelectric properties of PEDOT and its derivatives have received extensive attention and research. In particular, PEDOT: poly (styrene sulfonate) (PEDOT:PSS) has received the most attention because it can be doped through simple solution treatment techniques. However, the as-prepared PEDOT:PSS films tend to exhibit low thermoelectric properties, including electrical conductivity $< 1\,\mathrm{S\,cm^{-1}}$, Seebeck coefficient of $\approx 15\,\mathrm{\mu V\,K^{-1}}$, and PF $\approx 0.008\,\mathrm{\mu W\,m^{-1}\,K^{-2}}$ [42–44]. Therefore, further treatment of PEDOT with appropriate processes, especially the regulation of doping, can effectively improve the thermoelectric properties.

Secondary Doping of PEDOT:PSS: Secondary doping processes are usually referred to by adding chemicals to an aqueous solution of PEDOT:PSS, or treating PEDOT:PSS films with solvents, vapors, or solutions. This process can improve the conductivity of PEDOT:PSS by changing the composition of PEDOT:PSS and the morphology of the membrane without changing the doping level. Commonly used aqueous additives are compounds such as dimethyl sulfoxide (DMSO), ethylene glycol (EG), or poly(ethylene oxide) (PEO). For example, Gong et al. found that the addition of PEO and DMSO could not only improve the conductivity but also the Seebeck coefficient [45]. Thus, they observed an optimal PF of $157.4\,\mathrm{\mu W\,m^{-1}\,K^{-2}}$, conductivity of $1300\,\mathrm{S\,cm^{-1}}$, and Seebeck coefficient of $40\,\mathrm{\mu V\,K^{-1}}$. Posttreatment of PEDOT:PSS films is usually done with organic solvents and inorganic salts. For example, Chu et al. found that the PF of PEDOT:PSS films treated with formic acid was $80.6\,\mathrm{\mu W\,m^{-1}\,K^{-2}}$, the conductivity was $1900\,\mathrm{S\,cm^{-1}}$, and the Seebeck coefficient was $20.6\,\mathrm{\mu V\,K^{-1}}$ [46]. In addition, the inorganic salt solution prepared by adding $ZnCl_2$ into DMF was used to treat PEDOT:PSS films [47]. The conductivity and Seebeck coefficient were increased to more than $1400\,\mathrm{S\,cm^{-1}}$, and the best PF was $98.2\,\mathrm{\mu W\,m^{-1}\,K^{-2}}$. Because polar solvents and cations can weaken the coulomb interaction between PEDOT and PSS chain, leading to partial removal of PSS, the conformation of PEDOT chain changes, and the morphology of PEDOT:PSS film changes accordingly.

Dedoping of PEDOT:PSS: Generally, the as-prepared PEDOT:PSS has a high doping level and a low Seebeck coefficient. Dedoping can significantly improve the Seebeck coefficient, but dedoping can reduce conductivity. In addition, doping level can affect both energy band structure and state density (DOS), so it is necessary to find the optimal doping level. Dedoping can be done by chemical or electrochemical methods. For instance, PEDOT:PSS was dedoped electrochemically in organic electrochemical transistor [48]. An optimal PF of 23.5 μW m^{-1} K^{-2} was observed when the doping level was reduced from 33% to 14.5%. Chemical dedoping of PEDOT:PSS films is usually performed with reducing agents such as hydrazine (HZ), ammonium formate (AF), 1-ethyl-3-methylimidazole tetrafluoroborate ($EMIM-BF_4$), sodium borohydride ($NaBH_4$), and hydroiodic acid (HI).

Sequential Treatments of PEDOT:PSS: As previously mentioned, secondary doping can increase the electron conductivity of PEDOT:PSS, while dedoping can increase the Seebeck coefficient but decrease the electron conductivity. Then, if the two processes are combined in a certain order, the advantages of the two processes can be maximized, thus greatly improving the thermoelectric performance. Kim et al. first prepared PEDOT:PSS films by adding both *p*-toluenesulfonic acid and DMSO to PEDOT:PSS solution, and then treated the as-prepared films with hydrazine DMSO solution [49]. Interestingly, the Seebeck coefficient and electron conductivity of the PEDOT:PSS membrane prepared by this method are improved. The optimized PF is increased to 318.4 μW m^{-1} K^{-2} with the Seebeck coefficient of 50.4 μV K^{-1} and electrical conductivity of 1253.5 S cm^{-1}.

3.3.4 Carbon Nanotubes/Conductive Polymer Composites

CNTs are widely used in different research studies because of their excellent mechanical, thermoelectric properties. As a conductive filler, CNTs can be used to enhance the conductivity of polymer-based materials. The preparation of CNTs/conductive polymer composites is a common strategy to enhance the performance of organic TE materials. Single/double-walled CNTs (S/DWCNTs) are often used to improve the Seebeck coefficient and electrical conductivity of polymer-based composites, because the smaller bundles of S/DWCNTs and less aggregation can maximize the contact between the carbon nanobundles, thus forming a highly conductive π conjugate channel with the conducting polymer.

For CNTs/conducting polymer composites, the conductivity and Seebeck coefficient are directly related to the concentration of CNTs. There are two different explanations. The first is based on carrier concentration and mobility [50]. On the one hand, the increase in carrier concentration increases the conductivity but decreases the Seebeck coefficient. On the other hand, carrier mobility can improve both conductivity and Seebeck coefficient. The carrier concentration and mobility of the composites increase with the increase of CNTs content. In general, when the content of CNTs is low, the carrier concentration is low, and the Seebeck coefficient increases due to the increase in carrier mobility. However, with the increase of CNTs content, the Seebeck coefficient will be inhibited with the increase of carrier concentration. The second explanation is based on the energy-filtering effect of the polymer layer [51–53]. In CNTs/conductive polymer composites, the interfacial polymer layer acts as an energy barrier and preferentially allows higher energy carriers to pass through, this phenomenon is known as the energy filtering effect. Due to the energy filtering effect, higher energy carriers are "screened out," so that the average carrier energy in the composite is increased, so the Seebeck coefficient is further improved. The thickness of the polymer layer that exerts the energy-filtering effect determines the barrier. When the thickness is thin, the energy filtration effect is positively correlated with the thickness. When the thickness continues to increase beyond the optimal thickness, fewer carriers can pass the barrier, and the effect weakens, so does the Seebeck coefficient. The thickness of the polymer interface layer decreases with the increase of the concentration of CNTs. Therefore, a large Seebeck coefficient can be obtained by adjusting the appropriate content of CNTs to obtain the optimal polymer layer thickness.

The composition and structure of materials directly affect their properties, and this is also true for CNTs-based composites naturally, as described previously, the Seebeck coefficient can be directly affected by the regulation of polymer layer thickness. Differences in the preparation methods of CNTs will lead to differences in their structures and thermoelectric properties. The ultrasonic dispersion method commonly used in the experiment tends to shorten CNTs, resulting in reduced electrical conductivity. At the same time, the residue of the insulating surfactant can weaken and clog the contacts, further reducing the conductivity. Fortunately, the π–π interaction between the conductive polymer chain and CNTs may result in the ordered structure between the polymer chain and the nanotubes at the interface, leading to high carrier mobility, and increasing the conductivity and Seebeck coefficient. For composites, stacking different materials on top of each other, layer by layer, can yield surprisingly good performance. For example, Grunlan et al. composed PANI, graphene, PEDOT:PSS, and DWCNT layer by layer to form a thin film, then superimposed to form 80 layers of composite film, and finally obtained a maximum PF of 2710 $mW\,m^{-1}\,K^{-2}$ [54]. It is worth noting that both the conductivity and Seebeck coefficient increase with the number of stacked layers. The improved conductivity is dependent on the interfacial π–π interaction within the film, and the three-dimensional interconnect structure provides a more efficient way for electron transport. The high Seebeck coefficient is dependent on the energy filtering caused by the layered structure and the nanointerface between the components. Table 3.3 summarizes some representative CNTs/conductive polymer composites at room temperature.

In conclusion, CNTs/conductive polymer composites have the advantages of high conductivity of CNTs, low thermal conductivity of polymer, and high Seebeck coefficient obtained from structure optimization, so they are considered as thermoelectric materials with broad application prospects. Moreover, the treatment and doping processes mentioned in the previous section can be applied to CNTs/polymer composites to further improve the thermoelectric properties. In addition, CNTs matrix composites also have great advantages in the application of wearable thermoelectric devices due to their ease of processing and good flexibility.

3.3.5 Inorganic Semiconductive Nanomaterials/Polymer Composites

Inorganic semiconductor nanomaterials have a high Seebeck coefficient, which can be combined with conductive polymer materials with low thermal conductivity and good flexibility. Such composite thermoelectric materials can use conductive polymer (such as PEDOT and PVP) as the continuous phase and low-dimensional

Table 3.3 Thermoelectrical performance of some representative carbon nanotubes/conductive polymer composites at room temperature.

Materials	σ (S cm^{-1})	|S| (μV K^{-1})	PF (μW m^{-1} K^{-2})	k (W m^{-1} K^{-1})	ZT	References
DWCNT/graphene/PEDOT:PSS/PANI	1900	120	2710			[55]
SWCNT/PEI	3600	64	1500		0.5	[56]
SWCNT/(PEDOT:PSS)	1700	56	526	0.5	0.4	[57]
DWCNT/(PEDOT:PSS)/TCPP	960	70	470			[58]
SWCNTF/(PEDOT:PSS)/PVA	1000	40	160	0.2	0.3	[59]
SWCNT/PANI	1440	38.9	217	0.44	0.14	[60]
SWCNT/PS	1250	61	413	0.3	0.41	[61]

Table 3.4 Thermoelectrical performance of some representative inorganic semiconductive nanomaterial/polymer composites at room temperature.

Materials	σ (S cm^{-1})	|S| (μV K^{-1})	PF (μW m^{-1} K^{-2})	k (W m^{-1} K^{-1})	ZT	References
Te/PEDOT:PSS	19	163	70	0.2	0.1	[62]
PbTe/PEDOT	0.2	2500	1.4	—	—	[63]
Bi_2Te_3/PEDOT:PSS	60	150	131	0.56	0.08	[64]
SnSe/PEDOT	320	110	380	0.34	0.32	[65]
Sb_2Te_3/PEDOT	130	130	210	0.14	0.4	[66]
Bi_2Te_3/PEDOT	929	165	1350	0.7	0.58	[67]
$TiS_2/[(HA)_{0.08}(H_2O)_{0.22}(DMSO)_{0.03}]$	790	78	450	0.68	0.2	[68]

semiconductive nanomaterials (such as nanowires, nanorods, and nanoparticles) as the intensifying phase. At present, the main reinforcement materials including tellurium-(Te), selenium-(Se), titanium-(Ti) based materials. The design strategy of inorganic semiconductive nanomaterials/polymer composites can be started from two aspects: (i) interface optimization between continuous phase and reinforced phase; (ii) inorganic–organic superlattices. Table 3.4 summarizes some representative inorganic semiconductive nanomaterial/polymer composites at room temperature.

Interface optimization: The mixing of inorganic and organic components will produce a large amount of interface, and the interface properties will affect the TE parameters. The TE parameter is highly dependent on the interface area-to-volume ratio because the phase interface plays a key role in phonon scattering and carrier transport. In general, a large number of interfaces can increase phonon scattering and thus reduce thermal conductivity, which requires the enhanced phase to be uniformly mono-dispersed in the continuous phase to provide a larger interface surface. On the other hand, the continuous phase composed of conductive polymers has good conductivity and can provide favorable electron transport, while the monodisperse reinforced phase also facilitates the continuity of the polymer. The TE parameters of the composites can be optimized by controlling the particle size and dispersion of the intensifying phase material. The PEDOT/Bi_2Te_3 hybrid films manufactured using nanosphere lithography have monodisperse and periodic nanophases Bi_2Te_3, in which the size and spacing of the nanophases can be adjusted from tens to hundreds of nanometers, exhibiting an ultrahigh ZT value of ~0.58 at room temperature [67].

Inorganic–organic superlattices: In addition to dispersing inorganic materials in conducting polymers, inorganic materials can be stacked layer by layer with conducting polymers by electrochemical intercalation and solvent exchange to prepare inorganic–organic superlattice materials. Inorganic–organic superlattices can significantly enhance the power factor, even to the level of inorganic materials. The inclusion of inorganic components in polymers can enhance thermoelectric properties. Superlattices with electron transport but phonon-blocking structures can be used to produce flexible thermoelectric materials with excellent properties by regulating carrier concentrations.

3.4 Construction and Functionalization of TE Devices

Polymer-based thermoelectric materials with good flexibility have considerable application prospects in wearable devices that collect and convert low-grade heat. In addition, flexible thermoelectric modules based on polymer materials also show good performance in high-precision refrigeration. Thermoelectric devices are the core link to realize the leap from thermoelectric materials to thermoelectric conversion technology, and the device design integration covers the interdisciplinary science and technology issues such as thermodynamics, dynamics, thermal processing, interface physics and interface engineering, reliability design, optimization control theory, and so on. At present, compared with the rapid development of thermoelectric materials science, the research of thermoelectric devices is relatively weak, and most of the reported device research work is mainly confined to the preparation of π type elements to verify the material properties, and the energy conversion efficiency of practical thermoelectric components is mostly below 10%. At present, the measured conversion efficiency of the device is still far lower than the theoretical value based on the material ZT.

Construction: A conventional TE device consists of cuboids of n-type and p-type semiconductor material pair (thermoelement) that are connected electrically in series and thermally in parallel with a temperature gradient. Figure 3.3 shows the configuration of typical thermoelectric devices. When a thermocouple is connected

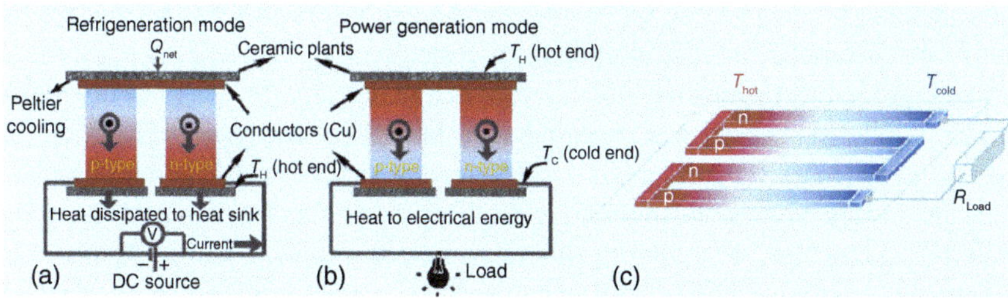

Figure 3.3 Basic principle and configuration of thermoelectric devices. (a) refrigeration mode; (b) power generation mode; (c) in-plane (printable) array of *p–n* couples. Source: Adapted from Kroon et al. [69]/Royal Society of Chemistry.

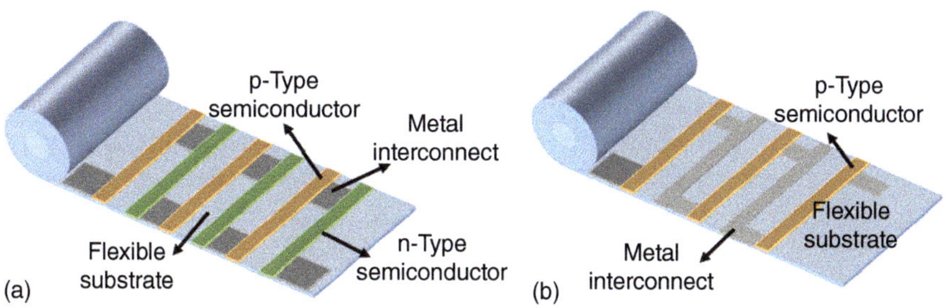

Figure 3.4 Schematic diagrams of two types of flexible TE devices consisting of (a) p–n legs and (b) p–p legs. Reproduce with permission. Source: Adapted from Chen et al. [70].

to an external load, as long as a certain temperature difference between the hot and cold sides is maintained due to an electrochemical potential, charge carriers begin to move around the closed circuit that provides electrical energy to the external load. However, this traditional geometric structure is not suitable for wearable TE devices based on organic TE materials processed in solution. Interestingly, a p–p legs configuration composed of one type of TE material is also proposed due to the rareness and poor environmental stability of n-type polymer-based TE materials (Figure 3.4). It should be noted that this configuration could incur heat loss because heat transfers from heat source to heat sink by thermal conduction from the metal wires, which causes degradation of the device performance.

According to the relationship between the direction of heat flow and the direction of basement extension, the reported organic thermoelectric devices mainly include vertical, horizontal, and inclined devices. Vertical devices have a typical π-like structure with heat flow direction perpendicular to the substrate. In order to meet the needs of flexible applications, the base materials of these organic devices are mostly flexible polymers and paper, which can be attached to aconite and other heat sources with different bending degrees to achieve efficient heat energy conversion. Although the device has the most widely used structure of inorganic thermoelectric materials, the thin film thickness of organic thermoelectric materials is usually low, which is not conducive to the establishment of large temperature differences, so many organic thermoelectric devices adopt a horizontal structure. Different from the control of film thickness, the horizontal film size is easy to control, so one can not only establish a large temperature difference on the unit device and improve the thermoelectric potential of a single group of devices but also form a highly integrated organic thermoelectric integrated device by curling, to meet the functional requirements of high voltage. Compared with the vertical and horizontal structures, the inclined structure is very similar to the fiber structure of the woven fabric and has outstanding advantages in the functional application of organic thermoelectric materials. Therefore, wearable organic thermoelectric devices can be prepared by using fiber as the base material through material composite and fine weaving strategies. This kind of device is easy to be prepared at a low cost and in large area and can effectively use the thermal energy released by the human body to generate electricity, showing the advantages of organic thermoelectric materials in the temperature difference of the room temperature.

Generally, there are three effective fabrication techniques for flexible thermoelectric (FTE) devices, such as the drop-casting technique, printing technique, and vacuum filtration technique. The drop-cast technique can fabricate free-standing films and requires further assembly processes to make a device, while the printing technique and vacuum filtration technique can directly fabricate devices. In addition, 3D printing technology has advantages in the rapid and streamlined manufacturing of CPs devices due to its programmable, simple, and flexible characteristics. Moreover, molding and lithography technologies have great advantages in high-precision manufacturing, with higher printing resolution than 3D printing, which can meet the needs of some ultra-fine electronic products composed of nanoscale patterns. Table 3.5 summarizes the construction and performance of conductive-polymer-based flexible TE devices, including the p-type and n-type TE elements used, substrates, units (couples), output voltages, output powers, output power densities, and applied temperature difference ΔT.

Functionalization: Conductive polymer is different from the rigid crystal structure of crystalline inorganic materials. The energy band or nanostructure engineering which is usually used to improve the performance of inorganic materials cannot be directly used to strengthen conductive polymer materials. Based on the unique molecular structure of CPs, different approaches are required, including inclusion, secondary inclusion, polymerization, post-processing, and hybridization. Although the performance parameters of polymer-based thermoelectric materials have been greatly improved, their ZTs values are still lower than those of inorganic materials, and they are difficult to operate at higher temperatures. In reality, the actual efficiency of TE devices is affected by a variety of factors, including heat loss and electrical and thermal contact resistance, resulting in a low-temperature difference between the hot and cold ends. For flexible/wearable TE devices, the actual efficiency achieved is quite low because the waste heat we utilize is of very poor quality. We tend to focus on their power output to meet the needs of portable electronics. The output power density pm is given as:

$$p_{\mathrm{m}} = \frac{P_{\mathrm{m}}}{WL} = \frac{PFt\Delta T^2}{2\Delta x^2} \tag{3.10}$$

where P_{m} is the identical peak output power, t denotes the internal resistance of TE system, ΔT is the temperature difference of the TE system and Δx denotes the length of a TE element.

Table 3.5 A summary of the construction and performance of conductive-polymer-based flexible devices.

p-type Material	n-type Material	Substrate	Unit	ΔT (K)	Output voltage (mV)	Output power (μW)	Output power density (μW cm^{-2})
PP-PEDOT (roll)	Bi_2Te_3	Polyimide	36	10	9–13	0.16	—
PEDOT:PSS	N-doped graphene	Plastic	—	10	3	1.92	0.024
PEDOT:PSS	—	Polyimide	14	12	2.9	—	1
PEDOT:PSS	Te	Polyimide	8	13	2.5	—	—
Sb_2Te_3 + PEDOT:PSS	Bi_2Te_3 + PEDOT:PSS	Polyimide	7	50	85.2	—	1220
—	Bi_2Te_3-epoxy	Polyimide	62	20	109	25	—
PEDOT:PSS/ SWCNTs	PEI-doped SWCNTs	Polyimide	6	50	28	0.22	—
PEDOT:PSS	Ni	Kapton	144	65	260	46	—
PEDOT:PSS	—	Paper	14	50	12	16.8	—
PEDOT:PSS	—	PES	5	25	2	—	—
CNT-coated activated carbon textile	—	PET + Fabric + Wire	2	31	0.000472	0.184	0.046
Au-doped CNT/ PANI webs	PEI-doped CNT webs	PANI + CNT Web	7	20	11	1.74	0.4
PEDOT:PSS	Constantan	Silicone rubber	10	30	12	0.38	0.1
$Sb_{37}Bi_2Te_{61}$	$Bi1_9Te_{75}Se_6$	Polyimide	112	13	1.41	26.3	18.63
PEDOT:PSS	—	Wadmal	—	65	7.28	1.2	0.04
SWCNT/WPU/ PVA	—	VHB tape	10	55	11	0.23	—
P_3HT doped by $AuCl_3$	—	3 M flip film	10	10	4	1.9	0.0185
Polyurethane/ EMIM: DCA	—	PTFE	—	0.4	10	—	0.00024
PEDOT:PSS	PEDOT:PSS/ $CuCl_2$	PET	10	4.5	1550	0.09	0.038
Ag_2Se/PVP	—	Polyimide	6	36.2	39.9	8.2	1.37
—	PVDF/ Ta_4SiTe_4	Polyimide	4	35.5	34	1.7	—

FTE devices typically have some flexibility in shape without breaking their structure, making them good candidates for wearable electronics for power generation and refrigeration. Therefore, high flexibility should be considered in addition to high performance and stability when designing FTE devices. In order to meet the requirements of wearability, TE devices should be designed with high flexibility, high TE performance, and a certain degree of comfort. At present, the design and manufacture of wearable TE devices is still in its infancy, and many efforts need to be made to further improve the performance of these wearable TE devices. There are two types of wearable TE devices, namely, film-based TE devices and fabric-based TE devices. The former usually consists of a two-dimensional polymer film, while the latter is more of a polymer wire woven or polymer-coated fabric material. Fabric-based wearable TE devices are either woven from polymer fibers or made from threads coated with polymers and their derivatives. Polymer fibers can be produced by the wet spinning process.

At present, the main challenges of organic thermoelectric devices focus on three aspects: accurate performance evaluation, basic strategies for device efficiency improvement, and unique function expansion. In terms of accurate performance evaluation, different parameters of thermoelectric materials require different device structures. In particular, thin-film thermal conductivity measurement has difficulties in film thickness, device size, and data fitting, resulting in the lack of widely accepted device performance evaluation methods and standards. In addition, the evaluation of the overall power generation, energy conversion efficiency, and Peltier refrigeration performance is still in its infancy. Overall, there is an urgent need to establish a widely accepted measurement and evaluation method system for organic thermoelectric devices. In the construction of high-performance devices, interface resistance, interface thermal resistance, material aggregation structure, and device structure all have important effects on the power output and energy conversion efficiency of devices, but these influence laws and optimization strategies lack of theoretical models and device optimization practice. In terms of thermoelectric process regulation and functional expansion, the key challenge is how to reveal the new effects and process regulation methods displayed in the thermoelectric energy conversion process of organic materials and construct new functional devices with advantages.

References

1 Toshima, N. (2017). *Synth. Met.* 225: 3.

2 Elsheikh, M.H., Shnawah, D.A., Sabri, M.F.M. et al. (2014). *Renewable Sustainable Energy Rev.* 30: 337.

3 Wang, Y., Yang, L., Shi, X.L. et al. (2019). *Adv. Mater.* 31: 29.

4 Serrano-Claurnarchirant, J.F., Culebras, M., Munoz-Espi, R. et al. (2020). *ACS Appl. Energy Mater.* 3: 861.

5 McGrail, B.T., Sehirlioglu, A., and Pentzer, E. (2015). *Angew. Chem. Int. Ed.* 54: 1710.

6 Hasan, M.N., Nafea, M., Nayan, N., and Ali, M.S.M. (2022). *Adv. Mater. Technol.* 7: 5.
7 Chen, G.R., Li, Y.Z., Bick, M., and Chen, J. (2020). *Chem. Rev.* 120: 3668.
8 Deng, L., Liu, Y., Zhang, Y. et al. (2022). *Adv. Funct. Mater.* 33: 2210770.
9 Russ, B., Glaudell, A., Urban, J.J. et al. (2016). *Nat. Rev. Mater.* 1: 10.
10 Matsui, H., Sato, T., Takahashi, T. et al. (2003). *Phys. Rev. Lett.* 90: 21.
11 Meng, J., Tang, Q., Zhou, L. et al. (2020). *Joule* 4: 953.
12 Miller, A. and Abrahams, E.J.P.R. (1960). *Phy. Rev.* 120: 745.
13 Lussem, B., Riede, M., and Leo, K. (2013). *Phys. Status Solidi A* 210: 9.
14 Zhou, D., Zhang, H., Zheng, H. et al. (2022). *Small* 18: 23.
15 Jacobs, I.E. and Moule, A.J. (2017). *Adv. Mater.* 29: 42.
16 Kanai, K., Akaike, K., Koyasu, K. et al. (2009). *Appl. Phys. A* 95: 309.
17 Patel, S.N., Glaudell, A.M., Peterson, K.A. et al. (2017). *Sci. Adv.* 3: 6.
18 Wang, S.H., Sun, H.D., Erdmann, T. et al. (2018). *Adv. Mater.* 30: 31.
19 Wei, P., Oh, J.H., Dong, G.F., and Bao, Z.N. (2010). *J. Am. Chem. Soc.* 132: 8852.
20 Cui, Y.T., Yan, J., Sun, Y. et al. (2018). *Sci. Bull.* 63: 814.
21 Bubnova, O., Khan, Z.U., Malti, A. et al. (2011). *Nat. Mater.* 10: 429.
22 Kim, G.H., Shao, L., Zhang, K., and Pipe, K.P. (2013). *Nat. Mater.* 12: 719.
23 Zhang, J.J., Song, G.J., Qiu, L. et al. (2018). *Macromol. Rapid Commun.* 39: 16.
24 Qu, S.Y., Yao, Q., Wang, L.M. et al. (2016). *NPG Asia Mater.* 8: e292.
25 Hynynen, J., Jarsvall, E., Kroon, R. et al. (2019). *ACS Macro Lett.* 8: 70.
26 Aich, R.B., Blouin, N., Bouchard, A., and Leclerc, M. (2009). *Chem. Mater.* 21: 751.
27 Jung, I.H., Hong, C.T., Lee, U.-H. et al. (2017). *Sci. Rep.* 7: 44704.
28 Ding, J.M., Liu, Z.T., Zhao, W.R. et al. (2019). *Angew. Chem. Int. Ed.* 58: 18994.
29 Sheng, P., Sun, Y.M., Jiao, F. et al. (2014). *Synth. Met.* 193: 1.
30 Schlitz, R.A., Brunetti, F.G., Glaudell, A.M. et al. (2014). *Adv. Mater.* 26: 2825.
31 Wang, Y., Nakano, M., Michinobu, T. et al. (2017). *Macromolecules* 50: 857.
32 Wang, S.H., Sun, H.D., Ail, U. et al. (2016). *Adv. Mater.* 28: 10764.
33 Shi, K., Zhang, F.J., Di, C.A. et al. (2015). *J. Am. Chem. Soc.* 137: 6979.
34 Yang, C.Y., Jin, W.L., Wang, J. et al. (2018). *Adv. Mater.* 30: 43.
35 Ma, H., Wang, X., Peng, Y. et al. (1810). *ACS Energy Lett.* 2019: 4.
36 Sun, Y.H., Qiu, L., Tang, L.P. et al. (2016). *Adv. Mater.* 28: 3351.
37 Sato, R., Kiyota, Y., Kadoya, T. et al. (2016). *RSC Adv.* 6: 41040.
38 Chaikin, P.M., Greene, R.L., Etemad, S., and Engler, E. (1976). *Phys. Rev. B.* 13: 1627.
39 Huang, D.Z., Yao, H.Y., Cui, Y.T. et al. (2017). *J. Am. Chem. Soc.* 139: 13013.
40 Huang, D.Z., Wang, C., Zou, Y. et al. (2016). *Angew. Chem. Int. Ed.* 55: 10672.
41 Yuan, D.F., Huang, D.Z., Rivero, S.M. et al. (2019). *Chem* 5: 964.
42 Luo, J.J., Billep, D., Blaudeck, T. et al. (2014). *J. Appl. Phys.* 115: 5.
43 Xiong, J.H., Jiang, F.X., Zhou, W.Q. et al. (2015). *RSC Adv.* 5: 60708.
44 Zhang, L., Deng, H., Liu, S.Y. et al. (2015). *RSC Adv.* 5: 105592.
45 Yi, C., Wilhite, A., Zhang, L. et al. (2015). *ACS Appl. Mater. Interfaces* 7: 8984.
46 Mengistie, D.A., Chen, C.H., Boopathi, K.M. et al. (2015). *ACS Appl. Mater. Interfaces* 7: 94.

47 Fan, Z., Du, D.H., Yu, Z.M. et al. (2016). *ACS Appl. Mater. Interfaces* 8: 23204.
48 Bubnova, O., Berggren, M., and Crispin, X. (2012). *J. Am. Chem. Soc.* 134: 16456.
49 Park, H., Lee, S.H., Kim, F.S. et al. (2014). *J. Mater. Chem. A* 2: 6532.
50 Wang, Y.Y., Cai, K.F., Shen, S., and Yao, X. (2015). *Synth. Met.* 209: 480.
51 Li, J., Huckleby, A.B., and Zhang, M. (2021). *J. Materiomics* 7: 51.
52 Wang, Y., Yu, C., Sheng, M. et al. (2018). *Adv. Mater. Interfaces* 5: 10.
53 Liang, Z., Boland, M.J., Butrouna, K. et al. (2017). *J. Mater. Chem. A* 5: 15891.
54 Cho, C., Wallace, K.L., Tzeng, P. et al. (2016). *Adv. Energy Mater.* 6: 7.
55 Chen, J., Wang, L., Gui, X. et al. (2017). *Carbon* 114: 1.
56 Zhou, W., Fan, Q., Zhang, Q. et al. (2017). *Nat. Commun.* 8: 14886.
57 Liu, S., Li, H., and He, C. (2019). *Carbon* 149: 25.
58 Moriarty, G.P., Briggs, K., Stevens, B. et al. (2013). *Energy Technol.* 1: 265.
59 Yu, C., Choi, K., Yin, L., and Grunlan, J.C. (2011). *ACS Nano* 5: 7885.
60 Wang, L., Yao, Q., Xiao, J. et al. (2016). *Chem. An Asian J.* 11: 1804.
61 Suemori, K., Watanabe, Y., and Hoshino, S. (2015). *Appl. Phys. Lett.* 106: 113902.
62 See, K.C., Feser, J.P., Chen, C.E. et al. (2010). *Nano Lett.* 10: 4664.
63 Wang, Y., Cai, K., and Yao, X. (2011). *ACS Appl. Mater. Interfaces* 3: 1163.
64 Niemela, J.P., Karttunen, A.J., and Karppinen, M. (2015). *J. Mater. Chem. C* 3: 10349.
65 Ju, H. and Kim, J. (2016). *ACS Nano* 10: 5730.
66 Zheng, W., Bi, P., Kang, H. et al. (2014). *Appl. Phys. Lett.* 105: 023901.
67 Wang, L., Zhang, Z., Liu, Y. et al. (2018). *Nat. Commun.* 9: 3817.
68 Wan, C., Gu, X., Dang, F. et al. (2015). *Nat. Mater.* 14: 622.
69 Kroon, R., Mengistie, D.A., Kiefer, D. et al. (2016). *Chem. Soc. Rev.* 45: 6147.
70 Chen, A., Madan, D., Wright, P.K., and Evans, J.W. (2011). *J. Micromech. Microeng.* 21: 10.

4

Liquid-Based Thermocells for Heat-To-Current Conversion

4.1 Introduction

Large amounts of low-grade heat (<100 °C) are emitted from industrial activities (such as power plants and data centers), are dispersed from the human body and microelectronics, and are spread in the environment (solar thermal and geothermal energy) (Figure 4.1). Low-grade heat is abundant, but due to its inherent low temperature and inefficiency, it is rarely used commercially. In order to do this, low-grade heat-to-electricity conversion systems must be efficient, scalable, and economical. A method of directly converting heat to electrical energy without the use of moving parts, audible noise, or carbon emissions is offered by thermoelectric (TE) technologies. Only a small number of technologies, such as TE energy conversion, thermal magnetic generators, thermal infiltration energy conversion, and thermal storage batteries, can currently be employed to make use of low-grade heat. They include conventional solid-state TE cells, which have undergone substantial study and are capable of directly converting low-grade heat into electrical energy in accordance with the Seebeck effect. The focus of conventional n-type or p-type electronic TE (e-TE) materials is often on semiconductors with a narrow bandgap, such as antimony telluride alloy or bismuth telluride alloy. As energy carriers, they make use of electrons or holes. Traditional solid-state TE cells, which are made of rare elements, are costly and often produce potential changes on the scale of 100–200 $mV\,K^{-1}$, which restricts the use of these devices for recovering low-grade heat.

As an alternative, liquid-state thermocells are promising for capturing low-grade heat due to their affordable, scalable, and adaptable characteristics (Figure 4.1). They do this by using the temperature dependence of electrochemical redox potentials (thermogalvanic effect) to generate electrical power. Thermocells are projected to have a cost-performance ratio of several dollars per watt, which is far less than that of conventional solid-state TE cells and on par with contemporary power-generating technologies. The high thermopower of thermocells on the scale of $mV\,K^{-1}$, which simplifies device design and integration to generate a high electric voltage despite a small temperature difference, is another benefit of using thermocells for low-grade heat harvesting.

Low-Grade Heat Harvesting: Materials, Devices, and Technologies, First Edition. Xiaogang Zhang.

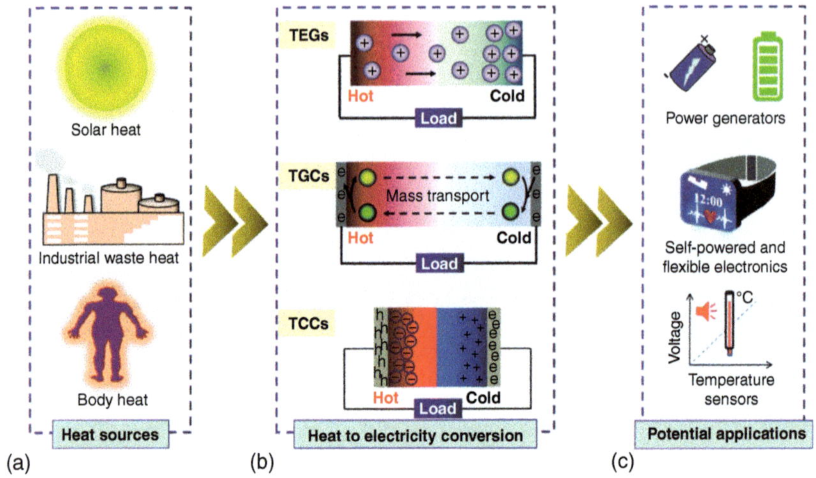

Figure 4.1 Heat to electricity conversion technologies of liquid-state thermocells. Source: Liu et al. [1]/Royal Society of Chemistry.

4.2 Basis of Thermocells Design

A liquid-state thermocell is an apparatus that produces potential difference by exploiting the impact of temperature gradient on the redox potential of electrochemically active substances. A redox pair, salt, and solvent typically make up the electrolyte, which is sandwiched between two electrodes in this system's basic structure. The temperature differential between the two electrodes is necessary to achieve TE conversion. A chemical reaction's pace and enthalpy change are intimately related to temperature. As a result, differing temperatures on the electrodes at the two ends result in distinct equilibrium constants for chemical reactions, resulting in different chemical equilibria at the cold and hot ends while tending to approach the same equilibrium at both ends. Therefore, the directional redox reaction and charge movement occur on the temperature gradient, forming a potential difference between the two electrodes, which enables the conversion of thermal energy to electrical energy (Figure 4.2) [2].

The thermodiffusion effect traces its origin to the Soret effect, first observed in the nineteenth century by Soret and Ludwig. In a thermodiffusible ionic thermoelectric (i-TE) material subjected to a ΔT, the ion concentration at two electrode surfaces is different because ions migrate from the hot to the cold side (Figure 4.3). Considering electronic neutrality, cationic and anionic ions coexist in i-TE materials. However, ions usually have different migration speed (diffusion rate) due to their different sizes, activation energies and Coulomb forces. The difference in diffusion rates between cations and anions leads to an imbalance in their respective ion concentrations. According to the Gouy–Chapman–Stern theory, a double-layer region consisting of a Helmholtz layer and a diffusion layer forms near the electrolyte/electrode interface, as shown in Figure 4.4. The difference in ion concentration leads to a

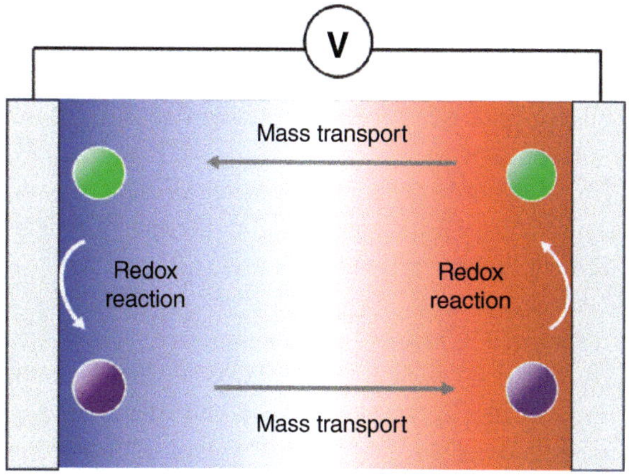

Figure 4.2 The illustration of liquid-state thermocell.

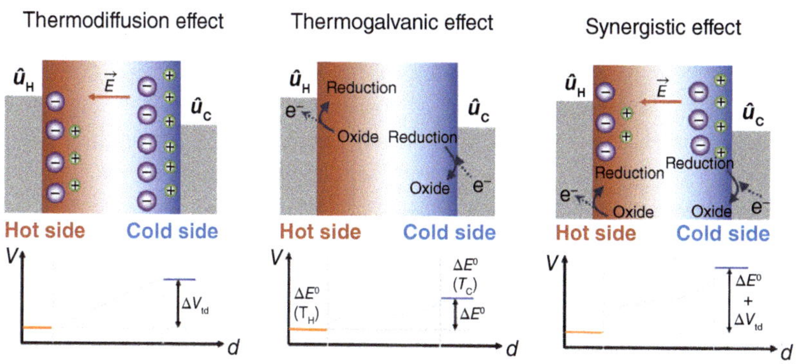

Figure 4.3 The working mechanism of liquid-state thermocell. Source: Adapted from Han et al. [3].

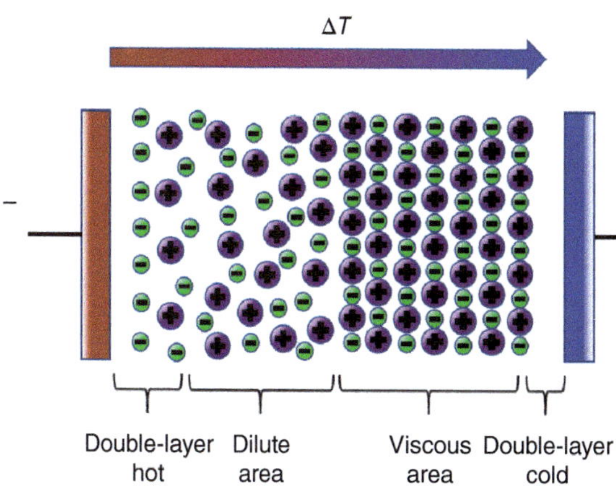

Figure 4.4 Internal structure of a thermodiffusive i-TE material in working mode.

viscous region near one electrode and a dilute region near the other electrode, which can generate a different chemical potential in their respective Helmholtz layers. The chemical potential difference between two Helmholtz layers inversely triggers an electric potential difference between two electrodes.

The main driving force of thermocells in the liquid state is the entropy transfer from the hot to the cold end, which consists of two parts: the transfer entropy SE, caused by a TE chemical reaction (thermogalvanic effect), and the transfer entropy SM, caused by material migration (Soret effect). The latter has a long reaction time and a small contribution potential difference (Figure 4.3). Therefore, the thermocell is mainly used for the conversion of heat into electricity according to the thermogalvanic effect. Unlike e-TE materials or thermodiffusive i-TE materials, the enthalpy in thermogalvanic i-TE materials changes discontinuously, so that the magnitude of the generated TE voltage is determined solely by the redox couples. For example, $Fe(CN_6)^{3-}/Fe(CN_6)^{4-}$-based i-TE materials usually have an S value greater than 2 mV K^{-1} due to the large Eastman entropy due to the resonant valence state of the $Fe(CN_6)^{3-/4-}$ ions. Figure 4.5 shows schematically the evolution of the chemical potential in a thermogalvanic process. Here, the thermogalvanic i-TE material is in an isothermal state, and the anode and cathode exhibit a balanced electric potential (E_0). When one of the electrodes is heated, the equilibrium is destroyed and the corresponding electric potential increases when an oxidation reaction occurs, or vice versa. When this system is connected to an external load, namely, in a discharge process, electrons flow from the anode to the cathode through the external circuit to balance the electrical potential difference.

Similar to solid-state TE cells, there are also the concepts of p-type and n-type thermocells. As shown in Figure 4.6, this is a typical p-type thermocell (oxidation on the hot side) with a negative TE voltage. In contrast, an n-type thermocell provides a positive thermopower.

It is well known that the performance of traditional solid-state TE cells is indirectly quantified by figure-of-merit ($Z = \frac{(S_i)^2 \times \sigma}{k}$) [5]. However, this value is not applicable for evaluating the performance of thermocell. One of the most important reasons is that the thermocell is a whole device. Its conductivity and thermal conductivity depends not only on the electrolyte and electrode but also

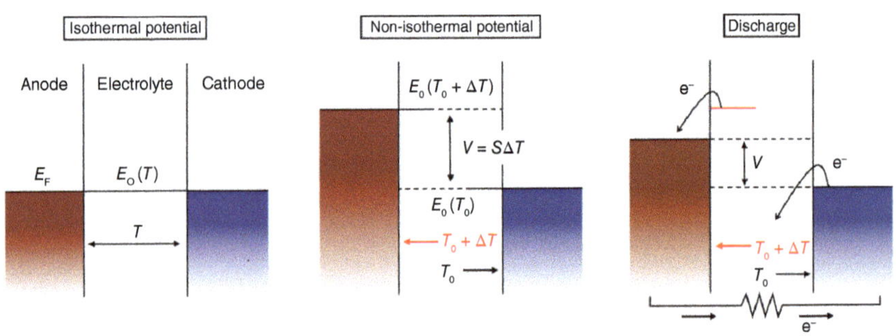

Figure 4.5 The electric potential evolution in a thermogalvanic i-TE material. Source: Adapted from Wang et al. [4].

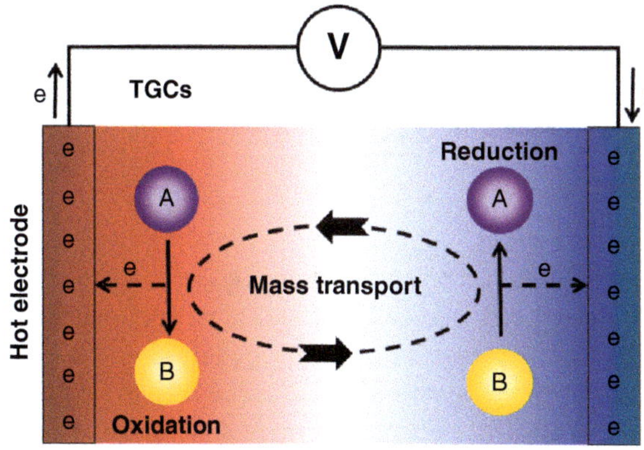

Figure 4.6 The configuration p-type and n-type thermocell.

Figure 4.7 Crucial parameters of thermocell.

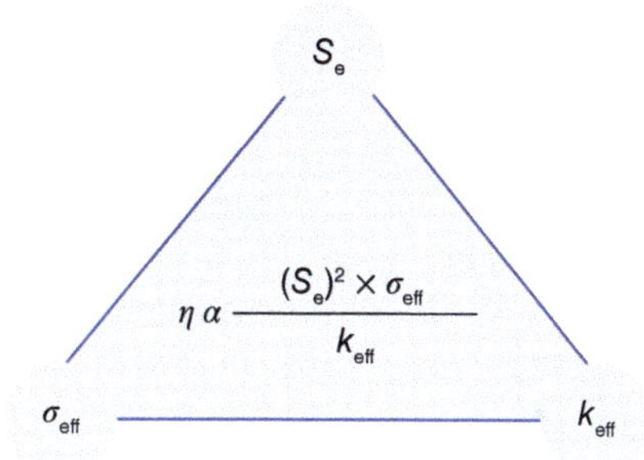

on the operating conditions. As known, redox reactions on the electrodes and mass transport in the electrolytes are the two critical processes for converting heat to electricity in liquid-state thermocells based on an electrochemical thermogalvanic effect. Alternatively, the performance of liquid-state thermocell is directly/indirectly quantified by Carnot-relative efficiency (η_r), thermopower (S_i), effective electrical conductivity (σ_{eff}), and effective thermal conductivity (k_{eff}) (Figure 4.7). Among them, the conversion efficiency of thermocells depends on the other three interdependent parameters.

For a temperature-dependent reversible redox reaction (A $\leftrightarrow$ ne$^-$ + B), the steady-state thermopower can be defined as the ratio of open circuit voltage (V_{OC}) to temperature difference (ΔT) as follow:

$$S_i = -\frac{\partial V}{\partial T} = -\frac{\Delta V}{\Delta T} = -\frac{V_{\text{H}} - V_{\text{C}}}{T_{\text{H}} - T_{\text{C}}} \tag{4.1}$$

S_i is also direct correlation with the reaction thermodynamics, which can be defined as temperature coefficient, expressed by:

$$S_{\text{tg}} = \frac{\Delta E}{nF} = \frac{\left(\widetilde{S_{\text{B}}} + S_{\text{B}}^*\right) - \left(\widetilde{S_{\text{A}}} + S_{\text{A}}^*\right) - n\overline{\overline{s}}_{\text{e}}}{nF} \tag{4.2}$$

where ΔG and ΔS are the Gibbs free energy and entropy change, $\widetilde{S}$ and S^* are the partial molar entropies and Eastman entropies of species A and B, and $\overline{\overline{s}}_{\text{e}}$ is the total transport entropy of electrons in external circuit. Because $\overline{\overline{s}}_e$ usually only contributes 1% in S_{tg}, it is usually ignored. Similarly, the relatively minor of S_A* and S_B* can also be ignored. Therefore, a simplification of formulas can be derived by ignoring such three parameters.

$$S_{tg} \cong \frac{\widetilde{S_B} - \widetilde{S_A}}{nF} \tag{4.3}$$

In thermocells without redox couples, anions and cations have obvious mobility under temperature gradient, which can be established by employing ionic electrolytes based on the Soret effect. The resulting ion concentration gradient destroys the overall neutrality of the electrolyte and generates a discontinuous thermal voltage due to the accumulation of ions at the electrode, forming a polarized electric double layer. Similarly, the thermodifusive thermal power of ionic species i can be defined as:

$$S_{td} = \frac{\partial E}{\partial T} = \frac{\sum \left(q_i n_i S_i^* D_i\right)}{\sum \left(q_i^2 n_i D_i\right)} \approx \left.\frac{\sum(\mu_{E,+}) - \sum(\mu_{E,-})}{\sum(\mu_{N,+}) - \sum(\mu_{N,-})}\right|_{n_+=n_-} \tag{4.4}$$

where n and D are concentration and diffusion coefficient of ionic species i, and $\mu_{E,+/-}$ and $\mu_{N,+/-}$ are thermal and electrical mobility of cations and anions, respectively.

As a result, the thermopower of thermocell combining thermodiffusive and thermogalvanic effect can be written as:

$$S_i = S_{tg} + \sum_i S_{td}(i) \tag{4.5}$$

Effective electrical conductivity can be comprehended from the reaction kinetics. It is a comprehensive electrical conductivity parameter, which is based on the internal resistance and charge transfer impedance of electrolyte, the charge transfer impedance of the electrode/electrolyte interface, the diffusion impedance in the bulk phase and the internal resistance of the electrode itself. Therefore, the electrochemical impedance spectroscopy is commonly adopted to characterize the resistances of thermocells.

Effective thermal conductivity combines the thermal conduction and convection of electrolytes under a temperature gradient, which is closely related to the diffusion coefficient and viscosity of electrolytes. Unlike solids, liquids are very mobile. Therefore, the thermal conductivity of fluids changes over a wide range when a temperature or force perturbation occurs. In general, the thermal conductivity of liquids is determined by conduction and convection. If the fluid is in an isothermal and flow-free state, thermal convection is negligible, and the thermal conductivity (intrinsic thermal conductivity) results from the intrinsic thermal conduction properties. However, if the fluid is in a non-isothermal state, such as liquid thermal cells with a large temperature difference, the effective thermal conductivity is much higher than its intrinsic thermal conductivity due to considerable convection formation. Therefore, the measurement method for thermal conductivity of liquid thermocells should be able to directly measure the effective thermal conductivity rather than the intrinsic thermal conductivity.

The other two parameters, namely, the temperature-insensitive maximum power density and Carnot relative efficiency, have attracted much attention. The former focuses on output performance, while the latter focuses on energy conversion efficiency. The relationship between these two parameters and the above parameters is shown in the formulas:

$$\frac{P_{\max}}{(\Delta T)^2} = 0.25 \times (S_i)^2 \times \sigma_{\text{eff}} \times \frac{A}{d} \tag{4.6}$$

$$\eta_r \approx \frac{0.25 \times (S_i)^2 \times \sigma_{\text{eff}}}{k_{\text{eff}}} \times T_h \tag{4.7}$$

where the terms of $(S_i)^2 \times \sigma_{\text{eff}}$ and $\frac{(S_i)^2 \times \sigma_{\text{eff}}}{k_{\text{eff}}}$ in formulas are termed as power factor (PF) and Figure-of-merit (Z), respectively, in which maximum power density with the temperature-independent feature is related to the configuration of thermcell, including the cross area of single electrode (A) and distance between two electrodes (d). It is worth noting that these measures still have limitations in comparing the advantages of devices. If designed thermocell get a large maximum power density and Carnot relative efficiency, the high PF and Z values are required accompanied by the huge thermopower and effective electrical conductivity with low effective thermal conductivity.

Apart from the above parameters, the ability of the thermocell to generate energy is best judged by its current output. Based on Ohm's law, the current is determined by simply connecting the variable external load resistor (R_{ext}) in series with the thermocell and measuring the potential (E) of the device, which can be described as follows:

$$I = \frac{E}{R_{\text{ext}}} \tag{4.8}$$

Figure 4.8 illustrates the mechanism of operation of the thermocell in the liquid state. As can be seen in Figure 4.8a, ion carriers migrate to the electrode surface at ΔT, which do not continuously pass through the external circuit. As a result, the ions accumulate and create an electric potential difference that acts like a capacitor (see Figure 4.8b). A single-charge/discharge cycle involves a four-stage process, with the corresponding voltage and ΔT. In the first stage, a specific ΔT is applied between the two electrodes so that ions gradually accumulate and produce an increasing TE

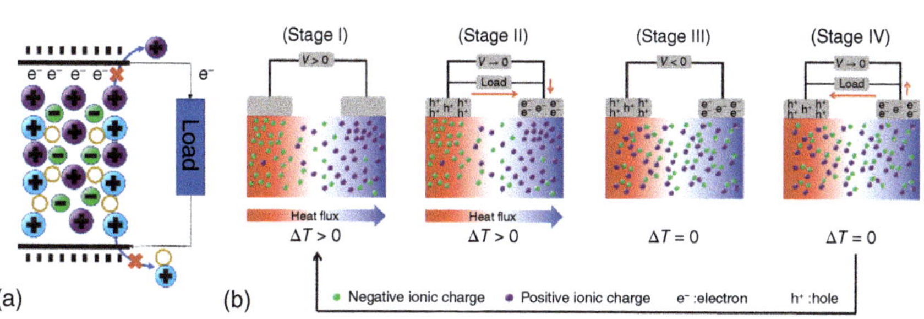

Figure 4.8 Schematic illustration of (a) working mechanism of thermocell, and (b) four stages of a thermocell capacitance cycle.

voltage. In the second stage, a closed circuit is established with an external resistor as the ΔT is kept constant. The voltage generated in the previous stage gradually drops to ≈0 V due to discharge. Since the accumulated ion amounts have reached a maximum concentration, the equilibrium voltage remains at ≈0 V thereafter. In the third stage, the ΔT is removed, and the external circuit is disconnected. As a result, the accumulated ions diffuse back to evenly distribute within the i-TE material. In the fourth stage, the external circuit is reconnected, and the electrons flow back to an electrically neutral state. An opposite-direction current compared with the second stage is thus released, which leads to zero voltage [6].

4.3 Engineering Strategies for Liquid-based Thermocells

4.3.1 Redox Couples

Redox couple is an essential component of liquid-state thermocell, while the thermal-to-electrical ratio and chemical stability are still never given any sort of satisfaction for application. The effectiveness of the redox couple can be evaluated by the thermogalvanic coefficient, which is determined by the entropy change in the TE chemistry process, including the redox species structure of partial molar entropy, the ionic solvation shell of Eastman entropy, and the electron to electrode transport. Here, Eastman entropy and partial molar entropy respect the interaction of an ion and its solvation shell with the solution and partial derivative of entropy with changes in the molar composition under constant temperature and pressure, respectively. Based on the difference of absolute entropy change between oxidant and reductant, the concept of p-type and n-type thermocell can be defined as positive thermopower and negative thermopower. In practical applications, there are other special factors to consider when selecting a suitable redox couple, such as solubility and chemical reversibility.

For example, the typical redox couple ferro/ferricyanide $Fe(CN)_6^{4-/3-}$ shows a huge entropy change of $-135.1\,J\,K^{-1}\,mol^{-1}$, resulting in a negative thermopower of $-1.4\,mV\,K^{-1}$, which was first reported by Quickenden and Vernon in 1986 using Pt electrodes in aqueous solution (Figure 4.9a) [7, 8]. Moreover, this value can be further improved to -2.9 and even $-4.2\,mV\,K^{-1}$ by adding additives into its aqueous electrolyte. Therefore, it has been widely utilized in liquid-state thermocell for energy harvesting. Iron (II/III) ($Fe^{2+/3+}$) and iodide/triiodide (I^-/I_3^-) are two other promising inorganic redox couples for thermocells, and both are n-type with positive thermopower (Figure 4.9b,c) [1]. The disadvantage for $Fe^{2+/3+}$ and I^-/I_3^- redox couples is their inferior thermopower with only 1.0 and $0.5\,mV\,K^{-1}$, respectively. Recently, both their thermopower have been adjusted over $2.0\,mV\,K^{-1}$ with great efforts. Apart from these intensively studied inorganic redox couples, other inorganic redox couples such as tin (IV/II) ($Sn^{4+/2+}$) and cobalt (II/III) ($Co^{2+/3+}$) are also investigated due to their large thermopower of 1.5 and $1.4\,mV\,K^{-1}$,

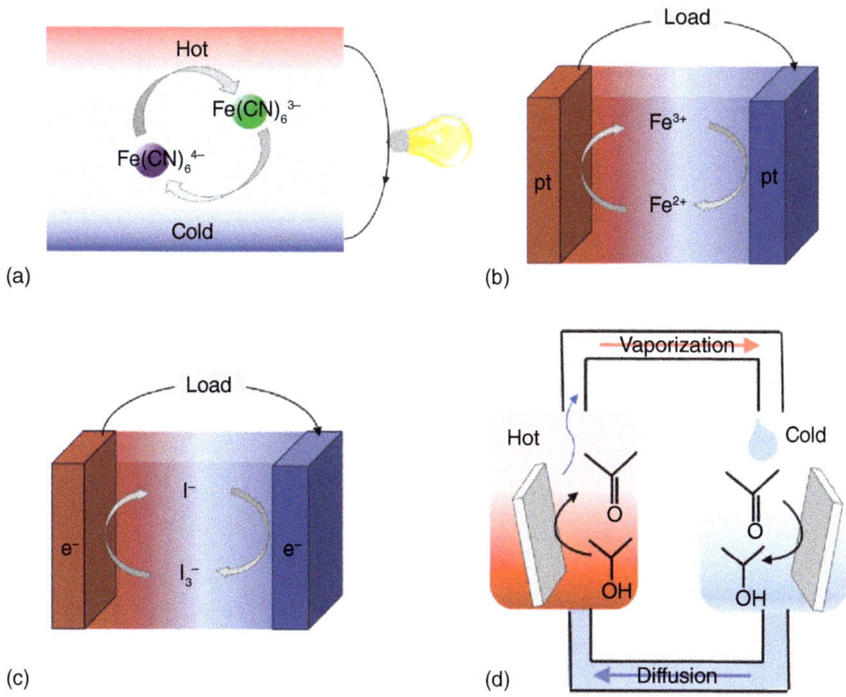

Figure 4.9 Inorganic redox couples of (a) $Fe(CN)_6^{4-/3-}$, (b) $Fe^{2+/3+}$, (c) I^-/I_3^-. (d) Organic redox couples.

respectively. Compared to inorganic redox couples, these redox couples possess higher thermopower (from −6.7 to −9.9 and 6.1 mV K^{-1}) (Figure 4.9d) [9, 10]. But their complexity in the synthesis and the process of phase transition somehow restrict their applications [11, 12].

Formula (4.2) describes the relationship between the thermopower of the thermocell and entropy change, where thermopower mainly depends on the entropy change of redox species during reaction in the reaction process. And entropy change is closely correlated with reaction thermodynamics. Namely, entropy change is related to the solvation-structural entropy difference (ΔS) and the concentration ratio difference (ΔC_r) between redox species, suggesting either improvement of ΔS or ΔC_r can efficiently boost thermopower. In general, ΔS value of redox species can be enhanced by regulating the interactions between the redox species and their solvent surroundings with special additives as shown in Figure 4.10a,b. The additive makes the redox species reorient and the complex solvation structure, thus improving the thermopower. For example, guanidine chloride is introduced into the aqueous thermocell containing $Fe(CN)_6^{4-/3-}$ redox couple to selectively induce $Fe(CN)_6^{4-}$ to crystallize on the cold side and dissolve on the hot side, thus establishing a persistent concentration gradient of $Fe(CN)_6^{4-/3-}$ between the hot side and the cold side (Figure 4.10c) [7]. In addition, Cu^{2+} can be induced

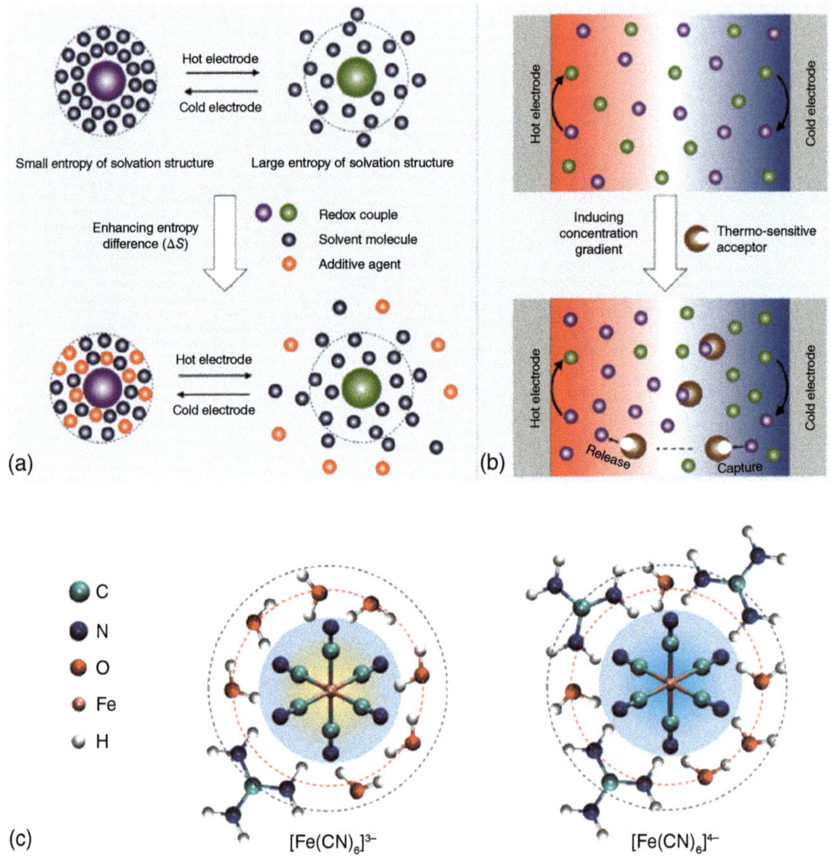

Figure 4.10 (a, b) The strategies of enhancing thermopower. Source: Duan et al. [2]. (c) the solvation structure of $Fe(CN)_6^{4-/3-}$. Source: Duan et al. [13]/Springer Nature/CC BY 4.0.

to crystallize on the cold side and dissolve on the hot side by adding $(NH_4)_2SO_4$, obtaining a thermopower of 1.66 mV K^{-1} [14]. Moreover, the introduction of temperature-sensitive additives such as nanogels can capture I_3^- at the hot side and then release I_3^- at the cold side, resulting in a concentration gradient of free I_3^- [15].

The entropy difference of reaction process can be expanded by host–guest chemistry, which has mainly focused on creating unique host–guest structures and their functions in the past few years. Typically, cyclodextrin (CD), a macrocyclic saccharide molecule, is added into the electrolyte to take advantage of the characteristics that the binding constant of host–guest complex is related to temperature selectively incorporate I_3^- into its hydrophobic cavity. The entropy of thermocell is decreased when I_3^- is captured by CD. Moreover, I_3^- is released from the CD cavity at the hot end, thus increasing entropy. Therefore, the net change of entropy and thermopower in thermocell is increased. It is worth noting that the increase of thermopower is estimated by calculating the concentration of free I_3^- at the cold electrode and the hot electrode [16] (Table 4.1).

Table 4.1 Review of typical redox couple in thermocell.

Year	Authors	Redox couple	Electrolytes	Electrode	Thermopower (mV K^{-1})
2020	J. Duan et al.	$Fe(CN)_6^{4-}$/ $Fe(CN)_6^{3-}$	Aqueous	Carbon paper	3.73
2019	G. Li et al.	$Fe(CN)_6^{4-}$/ $Fe(CN)_6^{3-}$	Aqueous	Anisotropic holey graphene aerogel	1.3
2018	J. Duan et al.	$Fe(CN)_6^{4-}$/ $Fe(CN)_6^{3-}$	Aqueous	Graphite	2.7
2018	J. Duan et al.	$Fe(CN)_6^{4-}$/ $Fe(CN)_6^{3-}$	Aqueous	Graphite	4.2
2016	H. Zhou et al.	I^-/I_3^-	Aqueous	Platinum	2
2019	J.H. Kim et al.	Fe^{2+}/Fe^{3+}	Aqueous	Platinum	1.76
2014	P.F. Salazar et al.	Co II/Co III	Ionic liquid	Platinum	1.6
2018	K. Kim et al.	Li/Li^+	Ionic liquid	Lithium	2.1
2010	R. Hu et al.	$Fe(CN)_6^{4-}$/ $Fe(CN)_6^{3-}$	Aqueous	MWCNTs	1.4
2012	T.J. Kang et al.	$Fe(CN)_6^{4-}$/ $Fe(CN)_6^{3-}$	Aqueous	SWCNTs	1.43
2013	M.S. Romano et al.	$Fe(CN)_6^{4-}$/ $Fe(CN)_6^{3-}$	Aqueous	SWNT–rGO composite	1.42
2016	H. Im et al.	$Fe(CN)_6^{4-}$/ $Fe(CN)_6^{3-}$	Aqueous	CNT aerogel sheet	1.43
2017	L. Zhang et al.	$Fe(CN)_6^{4-}$/ $Fe(CN)_6^{3-}$	Aqueous	Activated carbon cloth	1.43

4.3.2 Electrolyte

The diffusion coefficient and viscosity of the electrolyte have a great influence on the power density and thermal conductivity of thermocell. Aqueous electrolytes are widely used in thermal batteries due to their rapid ion diffusion and compatibility with many redox couples. Positive or negative thermopower can be obtained in aqueous electrolyte. Typically, Cu/Cu^{2+} and $Fe(CN)_6^{4-/3-}$ are +0.72 and −1.4 mV K^{-1}, respectively. On one issue, however, the application of aqueous-based thermocell is restricted by the low boiling point of water (100 °C), which stimulates the research of molecular solvents as electrolytes to enable the device to work at higher temperatures.

Electrolytes with unique properties can be produced by mixing various solvents. Typically, the TE potential of the deep eutectic solvent (DES) composed of choline chloride and ethylene glycol as the electrolyte of the $Fe(CN)_6^{3-/4-}$ thermocell is significantly higher than that in aqueous solution. This can be ascribed to the

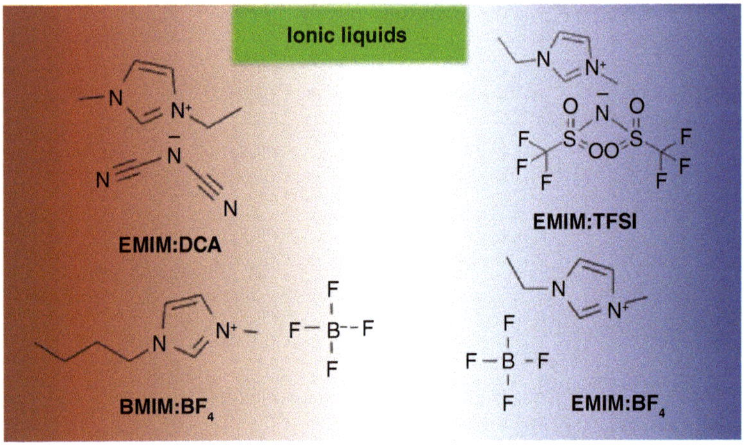

Figure 4.11 Chemical structures of typical ionic liquids.

disordered solvation structure in alcohol-based DES, and thus DES possesses a lower internal order, leading to the large entropy change of $Fe(CN)_6^{4-/3-}$ redox couple. In addition, DES electrolyte is different from water electrolyte with a low freezing point (−68 °C) and a high boiling point (210 °C), enabling the device to work in a wide range of temperatures.

Ionic liquids have been widely used in new energy storage device benefited from their negligible volatile, low flammability, high ion mobility, and wide electrochemical window. Moreover, the recent investigations have demonstrated the potential of (low) room-temperature ionic liquids-based thermocells. It is worth noting that ionic liquids act as energy carriers in thermocell and show huge thermopower in heat-to-current conversion in the field of TE generators. And the part of typical ionic liquids utilized in thermocell is shown in Figure 4.11.

The addition and type of salt in the electrolyte is also one of important factors affecting thermopower. Small cations such as Li^+ form strong bonds with water molecules, and $Fe(CN)_6^{3-/4-}$ is surrounded by ordered solvation cages (i.e. structure formation) [9, 17]. On the contrary, large cations such as Cs^+ form weak bonds with water molecules, and the solvation cage is not very orderly (i.e. structural fracture). It is possible that in the oxidation–reduction reaction, the loosely hydrated oxidation–reduction ion undergoes a large structural change in the solvation cage, thus increasing entropy change. The influence of ions on the surrounding structure of water clusters may be related to the Hofmeister series, which is an empirical rule developed based on the study of salting-out phenomenon of protein molecules. The typical sequence of cations in the Hofmeister series is: $(CH_3)_4N^+ > Cs^+ > Rb^+ > K^+ > Na^+ > Li^+ > Mg^{2+} > Ca^{2+}$. The magnitude of entropy change was found to decrease in the order of CsCl > RbCl > KCl > NaCl > LiCl, which matches the order of the Hofmeister series. This result confirms that the hydration structure around a redox ion is a deciding factor of entropy change, and the selection of a supporting electrolyte is also crucial in thermocell designs.

4.3.3 Electrode

In thermocells, as in other electrochemical devices, electrodes with high current density are one of the major limitations for energy conversion. Generally speaking, the exchange current density depends on the ion concentration and diffusion coefficient of the electrolyte, the charge transfer resistance at the electrode/electrolyte interface, and the electron transfer ability of the electrode material. Therefore, electrode materials with high catalytic surface, high conductivity, and large surface areas, such as platinum, carbon materials, and conductive polymers, are very promising. These properties ensure a high ion-accessible surface area, unobstructed ion diffusion channel, and good electron and thermal transfer [3].

Carbon textile or carbon cloth is a good candidate material with good electronic conductivity and intrinsic flexibility. Kim et al. reported the first electrode based on carbon fabric, in which carbon nanotubes (CNTs) were deposited on activated carbon fabric (ACT) by dip-coating method [10]. They proved that in the 0.4 M $K_{3/4}Fe(CN)_6$ electrolyte, the CNT-coated ACT exhibited lower charge transfer resistance (R_{ct}), mass transfer resistance (R_{mt}), and equivalent series resistance than the pure ACT, resulting in more than twice the thermal-electrochemical performance of the pure ACT in terms of normalized current density (139.15 vs. 62.32 mA m^{-2} K^{-1}) and normalized power density (48.86 vs. 21.9 μW m^{-2} K^{-1}). It is worth mentioning that a real flexible TGCs device is reported using this flexible electrode. Baughman et al. processed commercial carbon cloth products, which enhanced the exchange current density and output power of thermocell [18]. Commercial carbon cloth is modified by simple thermal annealing (1500 °C in argon) or soaking in an alkaline solution (2 M KOH), and then post-heat treatment (900 °C in argon). The thermal annealing proved that the conductivity could be increased from 2.6 to 160 S cm^{-2}, while the alkaline treatment increased the specific surface area (SSA) of the electrical activity, resulting in a porous structure, and increased the wettability of the electrolyte. Therefore, in the $K_3Fe(CN)_6/(NH_4)_4Fe(CN)_6$ electrolyte, the activated carbon cloth thin film electrode shows the highest power density (P_{max} = 1.80 mW m^{-2} K^{-1}).

Exploring flexible active materials is another effective way to realizc high-performance electrodes. In addition to excellent conductivity, large SSA, and good catalytic behavior, one-dimensional CNTs networks usually exhibit extraordinary mechanical properties. In the research group of Ray Baughman, CNT paper was made from multi-walled carbon nanotube (MWNT) network through vacuum filtration, which showed a high conductivity of 100 S cm^{-1} and a large SSA of 278 m^2 g^{-1} [19]. In 0.4 M $K_{3/4}Fe(CN)_6$ aqueous electrolyte, thermocells made of this MWNT paper provide a high output power of 0.38 mW m^{-2} K^{-1}, which is higher than that of platinum electrode (0.28 mW m^{-2} K^{-1}). After assembling the flexible MWNT film electrode into a thermocell device, it was wrapped on the cooling tube for encrgy collection, in which The OCV generated by the thermocell at a temperature difference of 15 °C is 21 mV, and the power density is 0.18 mW m^{-2} K^{-1}. However, the mass transfer of electrolyte ions is limited owing to the tortuosity of pure CNT electrode. To solve this problem, Chen et al. added reduced graphene oxide (rGO) to

SWNTs and achieved higher output power (0.46 $mW\ m^{-2}\ K^{-1}$). The improvement of performance is attributed to the increase of porosity and low curvature of composite membrane electrode, which is confirmed by the increased SSA and fast electron transfer kinetics. At the same time, the addition of two-dimensional rGO sheets induced the partial arrangement of the composite membrane, while maintaining good flexibility.

In addition to pure carbon-based materials, conductive polymers are another promising active electrode materials, because they have molecular porosity, inherent high specific surface area for small reactants, and are favorable ion transport conductors. Among them, the phase structure of poly(3,4-ethylenedioxythiophene)/polystyrene sulfonic acid (PEDOT/PSS) can be changed by impurities (such as ethylene glycol, diethylene glycol, dimethyl sulfoxide, or ionic liquid), and the polymer will show different electrochemical characteristics. Crispin et al. reported the electrode based on PEDOT/PSS in TGCs for the first time, in which PEDOT/PSS film was treated by secondary dopant dimethyl sulfoxide (DMSO). After DMSO treatment of PEDOT/PSS film with core-shell structure (conductive PEDOT/PSS core and insulating PSS shell), the excess hydrophilic PSS is separated into nano-domain, and an interconnected highly conductive PEDOT/PSS network is created. This infiltrating PEDOT/PSS network enhances the conductivity (from 0.4 to 805 $S\ cm^{-1}$, more than three orders of magnitude). Due to rapid electron transfer and good penetration of aqueous electrolyte, thermocell based on PEDOT/PSS exhibits a high output power of 0.45 $mW\ m^{-2}\ K^{-1}$, which is equivalent to developed carbon-based and platinum electrodes. PEDOT/PSS can also be introduced into the carbon system (graphene or CNT film), which can alleviate the problem of ion transport caused by the large circuity of CNT and the stacking of graphene electrodes. Chen et al. prepared a composite thin film electrode by combining PEDOT/PSS, edge functional graphene (EFG), and CNT through a simple drop-casting method. As dispersant and film-forming agent, dispersible PEDOT/PSS can provide more catalytic surfaces in the composite membrane in addition to providing ion transport pathway and increasing the porosity of EFG/CNT. Meanwhile, an additional laser etching process was adopted on the composite membrane, which promoted the penetration of electrolyte and improved the effective temperature difference between the cold and TE electrode/electrolyte interfaces. As a result, the instantaneous high power of 0.72 $mW\ m^{-2}\ K^{-1}$ was achieved in the viscous gel electrolyte containing 0.4 M $K_{3/4}FeCN_6$. They also studied the electrode in n-type electrolyte of PVA-$FeCl_{2/3}$ (1 M $FeCl_{2/3}$) system and proposed that pure PEDOT/PSS film showed excellent performance, $P_{max}/(\Delta T)_2$ is 0.38 $mW\ m^{-2}\ K^{-2}$, which exceeds all other composite materials, such as PEDOT/PSS-CNT, PEDOT/PSS-EFG, and PEDOT/PSS-EFG/CNT film. This is due to the high affinity of the negative charge on the redundant PSS in PEDOT/PSS for the multivalent cation, thus enhancing the transport of the cation in the polymer electrode (Table 4.2).

Rigid bulk materials with high catalytic or capacitive properties, such as platinum, copper, nickel, metal oxides, polypyrrole, and polyaniline, have been proven to possess excellent performance in TE devices, but are still not suitable for wearable and flexible applications. However, through nanoparticles, nanosheets, or nanowires to

Table 4.2 Review of typical electrode in thermocell.

Year	Authors	Electrode	Thermopower ($mV\,K^{-1}$)	Output power ($mW\,m^{-2}\,K^{-1}$)	Comments
2016	P. Yang et al.	Au/Cr	−1.21	0.012	The first report of gel electrolyte
2017	J. Wu et al.	Pt	−1.09	—	Excellent reversibility
2019	M. Russo et al.	Pt	−1.25	0.029	High concentration of electrolyte
2016	L. Jin et al.	Pt	−1.38	0.062	The use of gel electrolyte
2018	A. Taheri et al.	Pt	1.8	0.014	Electrode with thickness of 1 mm
2018	A. Taheri et al.	Pt	1.15	0.003	Quasi solid gel electrolyte
2016	P. Yang et al.	Au/Cr	1.02	0.033	Stretchable electrolyte
2020	Y. Liu et al.	CNT film	0.85	0.033	Elastic electrolyte
2018	J. Duan et al.	Graphite	−1.91	—	Reverse redox couple
2016	D. Zhao et al.	CNT	11.1	0.001	Polycationic electrolyte
2016	S.L. Kim et al.	PANI	8	0.576	Sulfonic acid polyanionic electrolyte
2017	F. Jiao et al.	Au	8.4	0.0635	Flexible and elastic
2019	B. Kim et al.	Au	16.2	7.637	High figure-of-merit
2020	Z.A. Akbar et al.	PANI	8.1	1.679	Self-healing electrolyte
2019	D. Zhao et al.	Au	14	0.117	High thermopower
2020	I.R Panneer-selvam et al.	CNT film	26	0.453	Mixed ionic electrolyte
2021	J. Xu et al.	Ag	34.5	1	Stretchable transparent electrolyte
2020	X. Wu et al.	Au	2.6	0.00593	p–n Type semiconductor connection setting
2020	S. Horike et al.	Ag	10	0.0116	Ionic gel
2019	T. Li et al.	Pt	24	1.152	Porous gel array
2018	K. Arpan et al.	PANI	1.21	0.00146	Redox gel electrolyte

carry out nanoscale engineering on their morphology, they can be incorporated into flexible, conductive, and porous main materials and have the opportunity to show their advantages in TE chemical properties. At the same time, it is also a good strategy to add platinum, copper, and bimetallic monoatomic dopants to the matrix material [20].

4.4 Direct and Indirect Thermocell Applications

Traditional thermosensors, such as thermopiles, are composed of inorganic alloys based on the e-TE effect. However, there are still straits of insufficient thermovoltage, material brittleness, and large-scale production obstacles. These drawbacks of traditional TE sensors restrict the detection limits and sensitivities. Recently, devices have been proposed as TE detectors owing to their giant ionic S values, high flexibility, and scale-up feasibility.

Theoretically, thermocell can generate electricity from all sources of heat, but its practical application is greatly limited. A key problem is that waste heat is usually generated as a by-product, such as waste heat from solar panels, data centers, and various electronic equipment. If the thermal battery is directly connected to these sources, its lower thermal conductivity will inhibit heat dissipation and further cause overheating of the heat source. Therefore, a typical static thermocell device is only suitable for obtaining low heat from sources that do not need heat dissipation (such as sunlight, human body, and exhaust pipe). Alternatively, the flowing thermocell may be an ideal choice, which can remove heat at the same time to keep the heat source within the working range.

In addition, the large-scale market application of thermocell is greatly limited by its low power density and energy conversion efficiency. The advancement of electrochemical and transport theory has promoted the development of high-performance thermocell. For example, due to the optimization of electrolytes and electrode materials, the power density of the $Fe(CN)6^{3-/4-}$ and the Carnot-relative energy conversion efficiency of the thermocell increased from 3.6×10^{-3} to $6.6\,W\,m^{-2}$ and 0.6% to 3.95%, respectively, because of the optimization of electrolyte and electrode materials.

However, due to the limited TE power and temperature difference, it is still challenging for a single battery to meet practical applications. In order to generate useful thermal voltage (>1 V) for practical applications, it is necessary to integrate multiple single cells into one configuration. Figure 4.12a shows the method of connecting a single device in series in two typical configurations. The connection at the top is named Z-shaped connection, which is only applicable to p-type or n-type, and the connection at the bottom is called π-shaped connection, which is applicable to p-type and n-type connections. Notably, compared with the significant improvement of p-type thermocell such as $Fe(CN)6^{3-/4-}$ systems, the n-type thermocell is significantly behind. Generally, n-type thermocell, including I^-/I_3^- and sulfate ($Fe^{2+/3+}$) oxidation and restoration pairs, is not sufficient owing to the slow reaction kinetics. Therefore, developing an efficient n-type thermocell is important for optimizing the

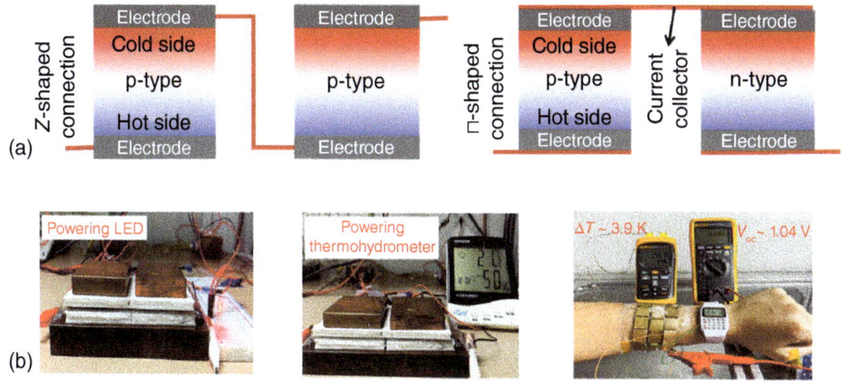

Figure 4.12 (a) Diagrammatic representation of a single cell connections. (b) Electricity generation and demonstration of using a TC-LTC module to power electronic devices. Source: Adapted from Li et al. [21].

device. As shown in Figure 4.12b, the designed device can supply power for various electronic devices at room temperature with a temperature difference of 50 °C, which uses p-type and n-type thermocells that combine electrical series connection and thermal parallel connection.

Thermal regenerative electrochemical cycle (TREC) is another indirect TE application based on temperature-dependent electrochemical potential. It has been used at high temperature (>500 °C) before, and its efficiency is up to 50% compared with Carnot engine. For cell reaction A + B → C + D (emission), temperature coefficient (α) defined as:

$$\alpha = \frac{\partial E}{\partial T} = -\frac{1}{n} \cdot \frac{\partial \Delta G}{\partial T} \tag{4.9}$$

where G is the partial molar Gibbs free energy. If α is negative, the system can discharge at low temperature and recharge at low voltage at high temperature. According to the temperature entropy diagram, TREC is a typical Ericsson loop, including isothermal charging and discharging and isobaric heating/cooling processes. The TE efficiency of TREC is defined as the ratio of the discharge electric energy required to charge the system to the input heat energy, which is mainly limited by heat recovery and internal resistance.

Another problem is the stability of the device under the wide temperature operation window. First of all, at high temperature, the equipment will face the high risk of liquid leakage, causing the volume of electrolyte to swell. In particular, when the temperature rises from 10 to 90 °C, the water-containing electrolyte shows 4% of the volume expansion; if the electrolyte is packaged in a rigid battery with a low-heat expansion coefficient, liquid leakage will occur. Filling the charging quality with stretching elastomers (such as polymailing silicone) or using gel-based electrolyte should be an effective way to solve the problem of liquid leakage. In addition, aqueous thermocell will also volatilize and boil at high temperatures. In this regard, ionic liquids with ignored volatility and good thermal stability as well as electrochemical stability are potential alternatives of water to extend the working temperature range (>100 °C) of the thermocell. Second, at low temperature, the device will face the problem of oxidation and restore substances and solvents, which may cause significantly reduced ion conductivity in the electrolyte. Adding low-temperature protective agents (such as inorganic salts and alcohol) to electrolyte is a promising method to expand the low-temperature operation window. In addition, for wearable gel base equipment, unstable contact between hydrogel and electrode is a challenge for the collection of human heat. The gel chemical bond is combined with an effective way to improve the stability of wearing heating battery devices on the electrode.

The increasing attention of the global energy crisis and the increasing attention to wearable devices stimulated the research on thermocell that use low-grade heat power generation. Due to the advancement of electrochemical and transport theory, compared with the Carnot engine, the development of thermocell with a thermal power of about 34 $mV\,K^{-1}$ and the energy conversion of about 11% has made a breakthrough. The equipment based on concept verification is feasible supplying power to various electronic equipment. However, because it is still low, it can only be eager for small-scale applications.

High-performance thermocells need to further improve the density of pupae and power density, while there are several key issues to be fully discussed (see unreasonable issues). One of the clear directions for further work should be to explore oxidative reduction puppets with better transmission and/or conduction performance, and electrolyte/electrode materials to optimize the Seebeck coefficient, ionic conductivity, and thermal conductivity, thereby increasing ZT value of these thermocells. However, engineering strategies, such as adding additives or membranes to the thermocell, or converting the direct thermal chemistry process into indirect processes, have also been proven to be valid. Further research in this field is expected to improve performance, thereby broadening the application scope of the thermocell.

Irrespective of serving as capacitors, generators, and sensors, the high-efficiency of devices is prominent. Insightful understanding surrounding the high-efficient device, the large-scale feasibility, the low-cost material innovation, the creation of novel battery architecture, the material flittering for different application scenarios, and the organization of biocompatible/wearable human body electronics are needed. Additional efforts should be paid to companion research on the minimization of contact resistance and the fabrication/functional modification of electrode materials as well.

Figure 4.13 illustrates a thermocell paper chip that is designed for remote alarming of early stage forest fires. After collecting the signal of ΔT, the telecommunication signal is transferred through a loop of signal conversion module, signal acquisition module, transmitter, and receiver in sequence. The whole system achieved signal transformation without relying on an additional power source, being low cost (≈\$0.4) and environment-friendly (paper matrix).

Unlike the commercialized e-TE materials, there have been technology gaps between experimental development and the practical application of i-TE systems. As the i-TE materials are lavish and interlock plenty of disciplines, the practical

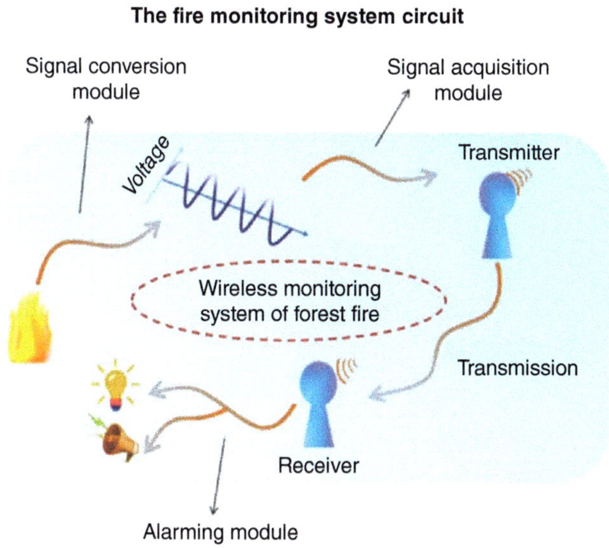

Figure 4.13 Schematic diagram of the communication system of the wireless remote fire alarm monitor. Source: Sun et al. [22]/John Wiley & Sons.

application of i-TE materials will also validate their prospects as a next-generation TE solution [1, 5].

References

1 Liu, Y., Cui, M., Ling, W. et al. (2022). *Energy Environ. Sci.* 15: 3670–3687.
2 Duan, J., Yu, B., Huang, L. et al. (2021). *Joule* 5: 768–779.
3 Han, C.-G., Qian, X., Li, Q. et al. (2020). *Science* 368: 1091–1098.
4 Wang, J., Feng, S.-P., Yang, Y. et al. (2015). *Nano Lett.* 15: 5784–5790.
5 Li, M., Hong, M., Dargusch, M. et al. (2021). *Trends Chem.* 3: 561–574.
6 Zhao, D., Wang, H., Khan, Z.U. et al. (2016). *Energy Environ. Sci.* 9: 1450–1457.
7 Yu, B., Duan, J., Cong, H. et al. (2020). *Science* 370: 342.
8 Kim, T., Lee, J.S., Lee, G. et al. (2017). *Nano Energy* 31: 160–167.
9 Ohtaki, H. and Radnai, T. (1993). *Chem. Rev.* 93: 1157–1204.
10 Im, H., Kim, T., Song, H. et al. (2016). *Nat. Commun.* 7: 10600.
11 Al-zubaidi, A., Ji, X., and Yu, J. (2017). *Sustainable Energy Fuels* 1: 1457–1474.
12 Guo, B., Miura, Y., and Hoshino, Y. (2021). *ACS Appl. Mater. Interfaces* 13: 32184–32192.
13 Duan, J., Feng, G., Yu, B. et al. (2018). *Nat. Commun.* 9: 5146.
14 Yu, B., Xiao, H., Zeng, Y. et al. (2022). *Nano Energy* 93: 106795.
15 Duan, J., Yu, B., Liu, K. et al. (2019). *Nano Energy* 57: 473–479.
16 Zhou, H., Yamada, T., and Kimizuka, N. (2016). *J. Am. Chem. Soc.* 138: 10502–10507.
17 Marcus, Y. (2009). *Chem. Rev.* 109: 1346–1370.

18 Hu, R., Cola, B.A., Haram, N. et al. (2010). *Nano Lett.* 10: 838–846.
19 Kang, T.J., Fang, S., Kozlov, M.E. et al. (2012). *Adv. Funct. Mater.* 22: 477–489.
20 Jiao, F., Naderi, A., Zhao, D. et al. (2017). *J. Mater. Chem. A* 5: 16883–16888.
21 Li, Y., Li, Q., Zhang, X. et al. (2022). *Adv. Energy Mater.* 12: 2103666.
22 Sun, S., Li, M., Shi, X.-L., and Chen, Z.-G. (2023). *Adv. Energy Mater.* https://doi.org/10.1002/aenm.202203692.

5

Thermosensitive Thermocells for Low-Grade Heat Harvesting

5.1 Introduction

Large amounts of low-grade heat (below 100 °C) are produced in natural sources and industries [1, 2]. For one thing, the low-grade heat from the geothermal activity and solar spectrum exceeds the annual energy consumption. For another, as much as 72% of the energy used in industrial production worldwide is wasted in the form of low-grade heat. Over the past few decades, many methods have been tried to capture this low-grade heat, but the efficiency of recovery remains low. More, commonly, the recovery rate of current thermoelectric devices is less than 1.4% due to material limitations; the efficiency of pyroelectric energy conversion is lower than 3.8%, and the temperature is unstable; due to the slow heat-driven ion diffusion mechanism, the recovery efficiency of low-grade heat in the liquid thermoelectric system is less than 5.7%.

Thermosensitive thermocells have high measuring accuracy, fast thermal response time, high mechanical strength, good voltage performance, long service life, and high measuring range (–200 to 1300 °C, under special circumstances can reach −270 to 2800 °C), it becomes an ideal low-grade heat conversion device. However, the correct use of thermosensitive thermocells is still a great challenge. The installation station and depth, insulation error, and thermal inertia and thermal resistance error have a certain influence on its accuracy. Low-grade heat signals are converted into thermal electromotive force signals to generate electricity. The conversion of low-grade heat in thermal electromotive force is partially important for the high-efficiency recovery of low-grade heat.

More recently, Duan and his team designed and manufactured thermosensitive-CsI_3-crystal-driven high-power I^-/I_3^- thermocells [3]. The present thermosensitive crystallization strategy to increase the Seebeck coefficient (S_e) from 0.51 to 1.2 $mV\,K^{-1}$, in 0.4 $mol\,l^{-1}$ I^-/I_3^- thermocells, leading to a maximum power density of 0.74 $W\,m^{-2}$. The additive of C_S^+ can selectively combine with I_3^- and from thermosensitive crystal CsI_3 at the cold side while releasing I_3^- at the hot side, which induces a concentration gradient of free I_3^-, thereby improving S_e. Besides, the CsI_3-driven I^-/I_3^- thermocell can assemble a high-performance device by integrating with a classic p-type cell. The thermosensitive crystallization strategy paves a new way to build high-power thermocells for low-grade heat harvesting.

The chapter systematically introduces thermosensitive thermocells from two aspects. First, the design principle and method of thermosensitive thermocells are introduced briefly, and the working principle and calculation method of cold-end compensation of thermosensitive thermocells are analyzed emphatically. Then, the performance test method and device integration technology of thermosensitive thermocells reported in recent years are discussed. Finally, the improvement of thermoelectric conversion performance of low-grade heat collection is summarized and discussed, the challenge of improving recovery efficiency is correctly understood and future development has prospected.

5.2 Design Principle and Method of Thermosensitive Thermocells

5.2.1 Working Principle of Thermosensitive Thermocells

Two different conductors or semiconductors A and B form a loop, and when the two ends are connected to each other, as long as the temperature of the two nodes is different, one end temperature is T, known as the working end or hot end, the other end temperature is T_0, known as the working end (also known as the reference end) or cold end (Figure 5.1). An electromotive force is generated in the loop, the direction and magnitude of which are related to the material of the conductor and the temperature of two contacts [4]. The accumulation of electric currents or charges as the temperature gradient moves from warmer to cooler regions is called the "thermoelectric effect." The loop of two conductions is called a "thermocouple," the two conductors are called a "thermal electrode," and the electromotive force produced is called "thermoelectric motive force." The thermoelectric electromotive force consists of two parts, one is the contact electromotive force of two conductors, and the other is thermoelectric electromotive force of a single conductor. The thermoelectric motive force in the circuit of the thermosensitive thermocells is only related to the conductor material and the temperature of two contacts and has nothing to do with the shape as well as the size of thermosensitive thermocells. After the two electrode materials of the thermosensitive theromcells are fixed, the thermoelectric motive

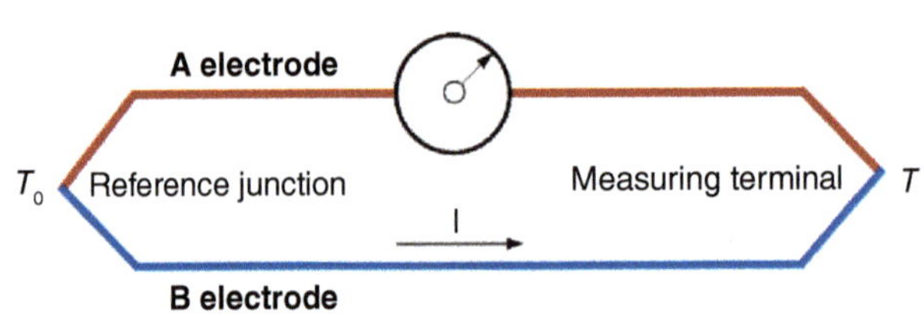

Figure 5.1 Schematic diagram of working principle of thermosensitive thermocells.

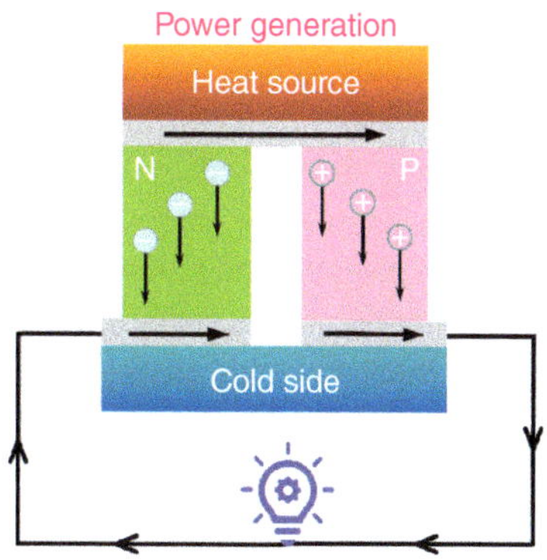

Figure 5.2 Schematic diagram of the Seebeck effect.

force is the functional difference between the T and T_0. The functional relation is as follows:

$$E_{AB}(t, t_0) = f(t) - f(t_0) \tag{5.1}$$

$$E_{AB}(t, t_0) = f(t) - C = \varphi(t) \tag{5.2}$$

The relation has been widely used in practical temperature measurement. Because T_0 is constant, the thermoelectric motive force produced by the thermosensitive thermocells only varies with the temperature at the hot end (measuring terminal), that is a certain thermoelectric motive force corresponds to a certain temperature. The purpose of temperature measurement can be measuring thermoelectric electromotive force (Figure 5.2).

Two material conductors with different components form a closed loop. When there is a temperature gradient at both ends, an electric current will be generated in the loop. At this time, there is an electromotive force between the two ends-thermoelectric motive force, which is known Seebeck effect. The calculation formula of the Seebeck effect potential difference is as follows:

$$V = \int_{T_1}^{T_2} (S_B(T) - S_A(T))dT \tag{5.3}$$

where S_A and S_B are Seebeck effect of two materials, respectively. If S_A and S_B do not change with temperature, the above equation can be expressed as follows:

$$V = (S_B - S_A)(T_2 - T_1) \tag{5.4}$$

The thermosensitive thermocells indexing table is made according to the functional relationship between the thermoelectric motive force and temperature. The indexing table is obtained under the free end temperature of 0 °C, and different thermosensitive thermocells have different indexing tables. When a third metal material is inserted into the thermosensitive thermocells loop, as long as the temperature of the two contacts of the material is the same, the thermoelectric potential generated by the thermosensitive thermocells will remain unchanged, that is, it will not be affected by the third metal access loop. In the thermosensitive thermocells, temperature measurement, can access the measuring instrument, measured thermoelectric motive force, and the temperature of the measured medium can be known. When the temperature of thermosensitive thermocells is measured, the temperature of its cold end is required to remain unchanged, and the size of its thermoelectric potential is proportional to the measured temperature. If the temperature of the cold end changes during measurement, the accuracy of measurement will be seriously affected. Therefore, it is necessary to take cold end compensation, that is, the cold end takes certain measures to compensate for the influence of the cold end temperature change on the thermal sensitive thermal battery.

Cold-end compensation of thermosensitive thermocells usually uses a bridge consisting of thermal resistance in series at the cold end. The three arms of the bridge are standard resistors, and one arm is made up of thermal resistors. When the temperature of the cold end changes, the thermoelectric potential generated by the thermosensitive thermocells will also change, and the thermal resistance value in the series. If the parameters are chosen well and the wiring is correct, the voltage generated by the bridge is exactly equal to the amount of thermoelectric potential that changes with temperature, and the total output voltage (potential) of the entire thermosensitive thermocells measuring loop responds accurately. The thermoelectric potential of thermosensitive thermocells is measured in terms of zero degrees of cold load. However, the instrument is usually measured at room temperature. Because the cold end is not zero, the thermoelectric potential difference decreases, resulting in inaccurate measurements and errors. Generally, the method of compensating wire is used. Indexing error: because the composition of the thermal electrode material does not meet the requirements and the uniformity of the material, the indexing error occurs between the hot spot property of the thermosensitive thermocells and the unified indexing table; compensation wire error: due to the difference thermoelectric of the compensation wire and the thermal motor material below 1000 °C, the error will be generated; error caused by reference temperature change: error caused by incomplete compensation when using compensation bridge for reference temperature compensation; and errors caused by changes in thermoelectric properties due to thermal electrode deterioration.

Cold-end compensation calculation method of thermosensitive thermocells is as follows:

(i) From millivolt to temperature: Measure the cold end temperature, convert it to the corresponding millivolt value, add it with the millivolt value of the thermocouple, and convert the temperature.
(ii) From temperature to millivolts: Measure the actual temperature and the cold-end temperature, convert them into millivolts, respectively, and subtract the millivolts to get the temperature.

5.2.2 Measurement Conditions of Thermosensitive Thermocells

In a closed loop composed of two conductors of different materials, different electron densities produce electron diffusion due to different materials, and electric potentials are generated after stability and equilibrium. When there is a gradient temperature at both ends, a current will be generated in the loop, resulting in thermoelectric electromotive force. The greater the temperature difference, the greater current in the loop. The temperature value is obtained after the thermoelectric motive force is measured. A thermocouple is actually an energy converter that converts heat into electricity.

Thermosensitive thermocells technical advantages: thermosensitive thermocells' temperature range is wide, the performance is relatively stable; high measurement precision, thermosensitive thermocells and measured object direct contact, not affected by the medium; the thermal response time is fast, and the thermocouple responds flexibly to the temperature change. The measurement range is large, thermosensitive thermocells from −40 to 1600 °C can be continuous temperature measurement. Thermosensitive thermocells have reliable performance and good mechanical strength, long service life, and convenient device. Two conductors or semiconductors of different materials are welded together to form a closed loop. When there is a temperature difference between conductors, an electromotive force is generated between them, thus creating a current of a magnitude in the loop. This phenomenon is called the thermoelectric effect. Thermosensitive thermocells is to use the thermoelectric effect to work.

5.2.3 The Basic Principle and Design Points of Thermosensitive Thermocells

Thermosensitive thermocells have been in widespread use since the early twentieth century. For many industrial and process-critical applications, understanding how thermocouples work can help you design circuits.

5.2.3.1 Working Principle of Thermosensitive Thermocells

When a voltage source is applied to a piece of wire, the current flows from the positive end to the negative end, and the wire heats up, causing some energy loss. The Seebeck effect, discovered by Tomas Seebeck in 1821, works in reverse: when a temperature gradient is applied to a piece of wire, an electric potential is created. That's the physical basis of thermosensitive thermocells.

$$\nabla V = S(T) \times \nabla T \tag{5.5}$$

where ∇T is the voltage gradient, ∇T is the temperature gradient, and $S(T)$ is the Seekbeck coefficient. The Seekbeck coefficient is material dependent and is also a function of temperature. The voltage between two different temperature points on a piece of wire is equal to the integral of the Seebeck coefficient function over temperature.

The thermosensitive thermocells consist of two different metals, and the Seebeck coefficient $S(T)$ of the wire is generally different. If a temperature in one metal can produce a voltage difference, why must two metals be used? Assume that the wire in Figure 5.3 is made of material "A." If the probe of A voltmeter is also made of material A, in theory, the voltmeter will not detect any voltage.

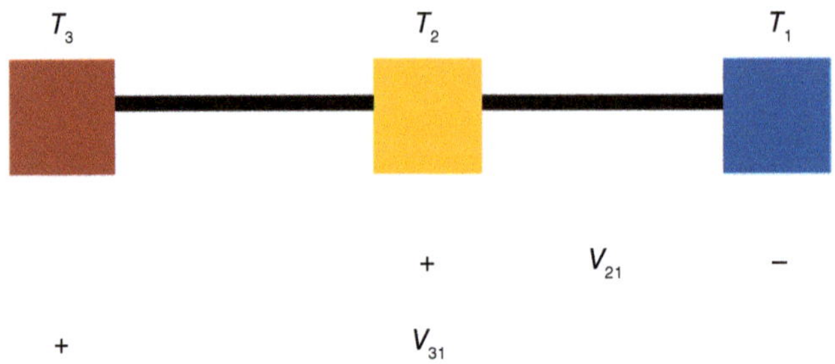

Figure 5.3 Diagram of a temperature gradient producing a voltage across a conductive metal.

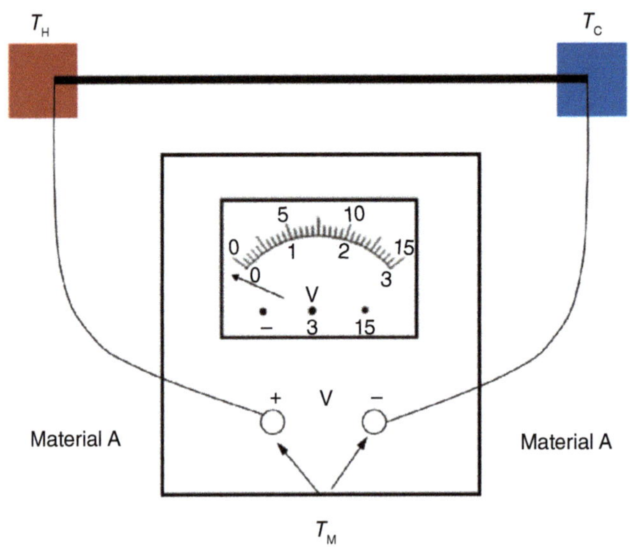

Figure 5.4 Voltage measurement connection diagram when probe and wire are of the same material.

The voltage formula in the loop is given as follows:

$$V = \int_{T_C}^{T_H} S_B(T) - S_A(T) dT \tag{5.6}$$

According to Figure 5.4 and Formula (5.6), assuming S_A, S_B, and the measured voltage is known, we still cannot calculate the temperature the T_H at the hot end unless we know the temperature T_C at the cold end. In the early stage of thermosensitive thermocells, a freezing with a temperature of 0 °C was used as a reference temperature (hence the term "cold end") because it was low cost, easy to implement and self-regulating [5]. The equivalent circuit diagram is shown in Figure 5.5.

Although we know the circuit reference temperature shown in Figure 5.5, it is not practical to obtain T_H by integrating. So, there are standard reference tables that

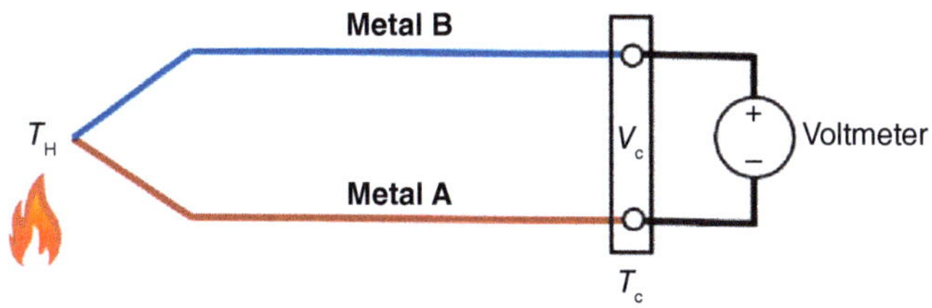

Figure 5.5 Reference temperature diagram of thermosensitive thermocells.

support the common thermosensitive thermocells types, all of which are drawn with 0 °C as the reference point.

5.2.3.2 The Hot End Temperature of Thermosensitive Thermocells is Calculated by CJC

All standard thermosensitive thermocells reference tables are obtained when the cold end is 0 °C. How to use the reference table to get the hot end temperature? Imagine extending the open end of the thermosensitive thermocells and connecting the imaginary end point to an anode with a temperature of 0 °C (Figure 5.6). If we can calculate the value of V_0, it is easy to obtain the corresponding hot end temperature using the reference table.

Determine V_0:

$$V_0 = \int_0^{T_C} S_B(T)dT + \int_{T_C}^{T_H} S_B(T)dT + \int_{T_H}^{T_C} S_A(T)dT + \int_{T_C}^{0} S_A(T)dT \tag{5.7}$$

Rearrange the above formula:

$$V_0 = \left(\int_{T_C}^{T_H} S_B(T)dT + \int_{T_H}^{T_C} S_A(T)dT\right) + \left(\int_{T_C}^{0} S_A(T)dT + \int_{0}^{T_C} S_B(T)dT\right) \tag{5.8}$$

$$V_0 = \int_{T_C}^{T_H} S_B(T)dT - S_A(T)dT + \int_{0}^{T_C} S_B(T)dT - S_A(T)dT \tag{5.9}$$

$$V_0 = \int_{0}^{T_H} S_B(T) - S_A(T)dT \tag{5.10}$$

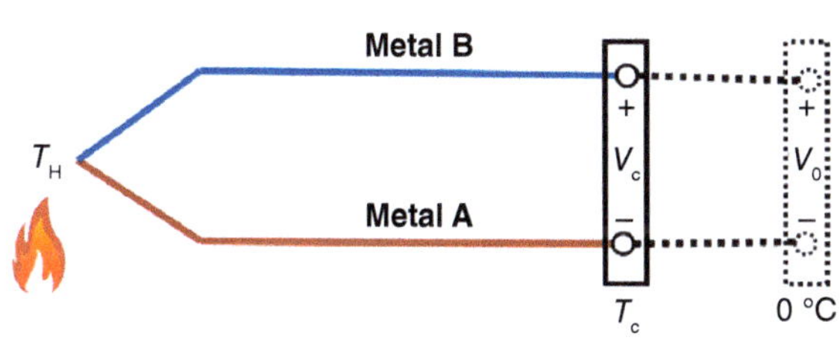

Figure 5.6 Schematic diagram of thermosensitive thermocells extension circuit.

The first term of Formula (5.9) is exactly the same as Figure 5.5. The equivalent voltage output is V_c, which is a known value because the temperature at the cold end is measured by the voltmeter. The second term is equivalent to the output of the thermosensitive thermocells when the temperature at the hot end is T_c and the temperature at the cold end is 0 °C. Since T_c is also measured by an independent temperature sensor, we can use Formula (5.9) for the second term corresponding to Seebeck voltage (V_i):

$$V_0 = V_C + V_i \tag{5.11}$$

Using the V_0 value, the corresponding temperature at T_H can be determined by the standard reference table. The process of calculating hot end temperature by cold end compensation is divided into the following step:

- Temperature sensor is used to measure cold end temperature (T_C);
- Measure the cold end temperature;
- T_c are converted to voltages by a standard reference table;

$$V_0 = V_C + V_i$$

- Convert V_0 to temperature T_H via a stand reference table.

5.2.3.3 Key Point of System Design of Thermosensitive Thermocells

So far, the above discussion is limited to the theoretical knowledge of the thermosensitive thermocells. To optimize the accuracy of real system, there are several considerations. Each component in the basic thermosensitive thermocells signal chain (Figure 5.7) will affect the conversion accuracy and must be carefully selected to minimize errors.

Starting from the left side of Figure 5.7, the thermosensitive thermocells are connected to the connectors on the system circuit board. The thermosensitive thermocells itself are also a sensor and may also have a source of error. Longer thermosensitive thermocells can easily pick up electrical noise from the surrounding environment, and shielding lines may be effective in reducing noise. The next component is the amplifier, which is important to have a high input impedance because the input impedance of the amplifier and the thermosensitive thermocells

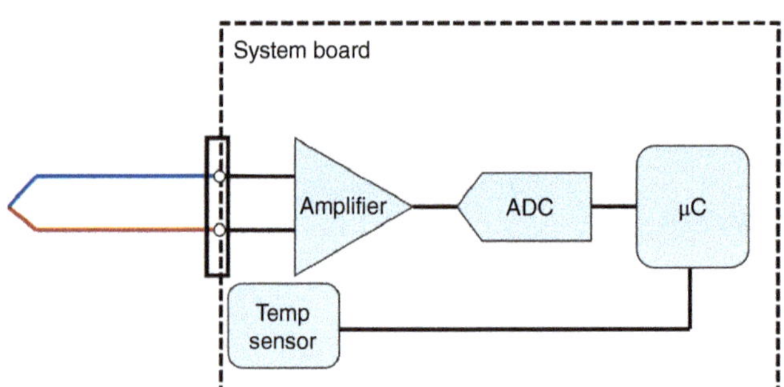

Figure 5.7 Schematic diagram of basic unit of thermosensitive thermocells system.

resistance form a voltage divider. The higher the input impedance of the amplifier, the smaller the error.

$$V_{\text{Amplifier input}} = V_{\text{Thermosensitive thermocell}} \times \frac{R_{\text{Amplifier impedance}}}{R_{\text{Amplifier impedance}} + R_{\text{Thermosensitive thermocell}}} \tag{5.12}$$

In addition, the amplifier amplifies the thermosensitive thermocells output, which is usually in the millivolt range. Although the amplifier's high closed-loop gain amplifies both signal and noise, adding a low-pass filter analog-to-digital converter (ADC) input eliminates most of the noise. Because the temperature does not change very quickly, ADC conversion rates for such applications are generally very low-perhaps sampling only a few times per second-so a low-pass filter is very effective. Finally, the onboard temperature sensor needs to be very close to the cold-end connector (ideally in contact with the end of the thermosensitive thermocells wire, but in many cases, conditions do not allow this) for the best cold-end temperature measurements. Any error in the cold-end measurement will be reflected in the hot-end thermometer.

5.2.3.4 The Basic Law of Thermosensitive Thermocells

(i) Law of mean conductor

A closed circuit is formed by welding both ends of the same homogeneous material (conductor or semiconductor). No matter how the conductor section is and how the temperature distribution is, no contact potential will be generated. The temperature difference potential will cancel out, and the total potential in the circuit will be zero.

It can be seen that the thermosensitive thermocells must have two different homogeneous conductors or semiconductors. If the hot electrode material is not uniform, due to the existence of temperature ladder, there will be a heating potential.

(ii) Law of intermediate conductor

The middle conductor (the third conductor) is connected in the thermosensitive thermocells loop. As long as the temperature at both ends of the middle conductor is the same, the introduction of the middle conductor has no effect on the potential of the thermosensitive thermocells loop. This is the law of intermediate conductors.

In line with the principles of intermediate conductors, practical applications of thermosensitive thermocells for temperature measurement often involve a configuration where the hot end is welded while the cold end remains open-circuited. The cold end is connected to the display instrument via connecting wires, thereby forming a comprehensive thermosensitive thermocells system.

There are concerns that the contact potential at the connection between the wire and the thermosensitive thermocells, when the cold end of the thermosensitive thermocells is connected to the meter to read the mV value, will cause additional errors in the measurement. According to this law, there is no error.

(iii) Law of intermediate temperature
The thermoelectric potential between the two connecting points of the thermosensitive thermocells loop (temperature T, T_0) is equal to the algebraic sum of the thermoelectric potential at temperature T, T_n and the thermoelectric potential at temperature T_n, T_0. T_n is called the intermediate temperature.
Application: because the relationship between the thermocouple E^{-1} is usually nonlinear, when the cold end temperature is not 0 °C, the actual thermoelectric potential E of the known loop cannot be used, which needs to be corrected according to the law of intermediate temperature.

(iv) Refer to the electrode law
A high-purity platinum wire is used as the standard electrode. Suppose that the positive and negative electrodes of nickel-chromium alloy/nickel-silicon alloy are paired with the standard electrode, and their values add up to the values of the nickel-chromium alloy/nickel-silicon alloy electrode.

5.2.3.5 Thermosensitive Thermocells Requirements for Thermal Electrode Materials

Thermosensitive thermocells electrode materials are a kind of new functional materials with the cross-coupled thermoelectric transport property. Using the properties of thermoelectric materials, heat energy and electric energy can be directly converted into each other. The Seebeck effect discovered in 1823 and the Partig effect discovered in 134 provided theoretical basis for the application of thermoelectric energy converters and thermoelectric refrigeration.

According to the principle of thermosensitive thermocells, it seems that any two different conductor materials can be used as a thermal electrode to form a thermosensitive thermocells, but in fact, it is not so, because there are certain requirements for thermosensitive thermocells thermal electrode materials.

- The thermoelectric potential should be large enough, and the relationship with temperature should be linear or nearly linear;
- In the use of temperature range, the physical and chemical properties are stable;
- Stable thermoelectric performance, easy to reproduce, good interchangeability of similar thermosensitive thermocells;
- High conductivity, small resistance temperature coefficient; and
- Convenient processing, cheap price.

However, materials that fully meet the above requirements are difficult to find. Therefore, only according to the temperature of the high and low, choose the appropriate thermoelectric materials.

The thermocouple composed of precious metals and alloys becomes the precious thermosensitive thermocells, collectively referred to as the base metal thermosensitive thermocells. Compared with noble metal thermosensitive thermocells, base metal thermosensitive thermocells have high sensitivity, approximate linear thermoelectric characteristics, and can be used in reducing atmospheres. But the oxidation resistance and corrosion resistance are not good, the uniformity of hot electrode is poor, the high-temperature stability is poor, and the life is short.

In common combinations of thermosensitive thermocells materials, at least one electrode is made of a transition metal or alloy, because their thermoelectric potential rate is much higher than that of a simple metal at any temperature. In the past, a large number of theoretical and experimental studies also focused on this, especially nickel, iron, platinum group metals, and refractory elements, and created a number of famous thermocouple materials, such as nickel-aluminum, nickel-copper, iron–copper, high temperature for platinum–rhodium–platinum, iridium rhodium–iridium, tungsten, molybdenum, and tungsten–rhenium [6]. In order to create new materials and improve traditional materials, it is obviously very important and basic significance to study this kind of metal materials in detail.

In addition, with the development of cryogenic engineering and superconductivity technology, the thermosensitive thermocell suitable for low temperature is urgently needed. Thermosensitive thermocells for cryogenic use of liquid hydrogen and helium are difficult to develop for two reasons. One is that the thermoelectric potential rates of general metals become very small in the deep cold range. Second, in the deep cryogenic domain below 77 K (liquid nitrogen), impurities have a great effect on the thermoelectric rate of the metal. Therefore, it is difficult to copy the thermoelectric properties of low-temperature thermosensitive thermocell materials. Sometimes even the presence of trace amount of impurities will make the thermoelectric potential rate of pure copper not only in terms of magnitude but also in terms of symbols. As early as the 1940s, during the study of the resistance properties of metals and alloys, scientists discovered the presence of certain unique alloys that exhibited extremely low resistance at low-temperatures. Within the realm of deep low-temperatures, these alloys demonstrated a thermoelectric potential rate significantly higher—ranging from a hundred to a thousand times—than that of typical metal alloys. The emergence of these special thin alloys has brought forth promising possibilities for researchers in their quest for ideal materials for deep low-temperature thermosensitive thermocell. In recent years, physicists refer to these special alloys as dilute "magnetic" alloys based on experimental results and theories. Unlike the magnetic systems of Fe, Ni, Mn, and Cr in the ordinary sense, the alloys involved here just exclude those showing magnetic transformations. Specifically speaking, the alloy system referred to here is a thin alloy with non-transition metals as solvents, transition metals or rare earth, and actinide metals as solutes, and the solute concentration is very low, only a few atomic percentages to a few hundredths of an atomic percentage or less. It is called a thin alloy for two reasons: first, it has been observed that the properties shown in many such alloys are consistent with the properties shown in alloys and with the concept of the existence of a local magnetic moment in alloys; secondly, it has been gradually recognized that spin fluctuations have an important effect on electron transport properties in many of these alloys. In fact, these dilute magnetic alloys have become the mainstream of the development of low-temperature thermosensitive thermocells materials today.

To develop a suitable new material, it is necessary to consider not only the thermoelectric property but also many metallurgical factors, metallic problems, physical and chemical properties, possible use of advanced purification techniques, and material manufacturing processes. Due to the segregation of chemical composition, impurity elements and physical defects may be produced in the process of material manufacturing, resulting in the inherent nonuniformity of materials. In the subsequent use process, the preferred evaporation, oxidation, diffusion, dissolubilization or external immersion of chemical components, or impurities will also cause inhomogeneity. When the two are in the temperature gradient field, the thermoelectric properties will be unstable, requiring material researchers to seek solutions. In addition, the material's plasticity and workability, mechanical and thermal strength, oxidation resistance and chemical inertness, and resources and price are also considered. In general, possible idealized purification techniques are used to ensure that materials have suitable high-purity states. A compact and continuous protective film is formed on the surface of the alloy by adding some trace alloying elements, in which oxygen and alloying elements are diffused. Select the alloy which is not sensitive to the composition fluctuation and has strong stain resistance instead of the pure metal; pay attention to the chemical compatibility between wire, insulator, and outer bushing when making armored thermocouple; and adopting proper heat treatment system, proper thermosensitive thermocells structure, correct installation, accurate division and timely calibration are all effective measures and methods to obtain high quality and high stability materials and realize accurate and reliable thermosensitive thermocells assembly.

5.3 Performance Test Method and Device Integration Technology of Thermosensitive Thermocells

Thermosensitive thermocells are mainly composed of cold and hot electrodes and electrolytes and rely on the temperature difference between the two electrodes and the electrolyte ion concentration difference to work (electrode temperature difference and ion concentration difference are the key factors to determine the Seebeck coefficient). However, in a closed system, the ion concentration gradient is a thermodynamic nonequilibrium state, which will spontaneously transit to a thermodynamic equilibrium state with uniform ion distribution through diffusion, so the ion concentration gradient basically returns to zero in a stable state, which means that the final thermochemical battery can only rely on the temperature gradient to maintain its operation, which will inevitably weaken the performance of the battery. The simultaneous optimization of the Seebeck coefficient and thermal conductivity is realized by inducing a stable and sustainable ion concentration gradient with thermosensitive crystal materials. By introducing a thermosensitive crystal material, a stable ion concentration gradient can be formed at the hot and cold ends of the device in the presence of a temperature difference, and the simultaneous synergistic optimization of the Seebeck coefficient and the thermal conductivity is realized. Thermosensitive thermocells are the most basic unit to realize the function of thermoelectric energy conversion, which is generally composed of p-type and n-type thermoelectric materials connected in series/parallel by electrodes, electrically insulating substrates, and so on. The figure of merit of the thermoelectric material not only determines the theoretical maximum efficiency of the device but also the topology of the device (geometry, size, connection, current and heat flow coupling matching, etc.) and heterointerface (electrode and thermoelectric material, electrode and insulating substrate, etc.). Such structural factors seriously affect the actual energy conversion efficiency, power density, reliability, and other service characteristics of the device. The diversity of applications of thermoelectric conversion technology puts forward diversified requirements for the structure and performance of devices, and different application objectives and service environments put forward different requirements for the topological structure and physical and chemical properties of devices. For example, thermoelectric devices for power generation applications, high temperature, large temperature difference, thermal-mechanical coupling, and other complex and harsh service environments put forward strict requirements for the reliability of devices for environmental energy collection applications, and the small temperature difference of service environment requires devices to have high response sensitivity to temperature. The device is required to have good flexibility and long-term stability in the flexible state. Maximizing the performance of thermosensitive materials, improving the actual conversion efficiency of devices, and meeting the core requirements of practical applications for their functions and structures are the core scientific and technical issues of device optimization design and integrated manufacturing technology.

5.3.1 Electrode Material Selection and Electrode-Connected Technology of Thermosensitive Thermocells

In the design and integration of thermosensitive thermocells, the connection between electrodes and thermoelectric materials, especially the design and connection of high-temperature electrodes of power generation devices, is the key technology of device integration. The efficiency, reliability, and cycle life of the device are directly affected by the physical properties of the electrode material itself (thermal conductivity, electrical conductivity, thermal expansion coefficient, etc.), its matching with the thermoelectric material, and the bonding state between the electrode and the thermoelectric material (bonding strength, interface resistance, interface thermal resistance, interface high temperature and chemical stability, etc.).

The selection of electrode materials mainly follows the following principles:

(i) The electrode material has high electrical conductivity and thermal conductivity to reduce energy loss.
(ii) The expansion coefficient of the electrode material should be as close as possible to that of the thermoelectric material connected with it, so as to avoid stress concentration and cracking caused by the mismatch of the thermal expansion coefficient. Generally, the difference between the thermal expansion coefficients of the electrode and the thermoelectric material needs to be controlled within 20%.
(iii) The bonding strength of the interface between the electrode and the thermoelectric material is high, and the contact resistance and the contact thermal resistance are low.

(iv) Within the temperature range, there is no serious mutual diffusion or reaction between the electrode and thermoelectric material, so as to avoid damage to the thermoelectric material and the electrode has certain oxidation resistance and higher temperature stability.

(v) The combination process of electrode and material is simple. In general, most metal materials are difficult to meet the above requirements simultaneously, especially high bonding strength, low interfacial resistance, and high-temperature stability of the interface are difficult to achieve at the same time, so multilayer interface layer is usually introduced to achieve the above multiple contradictory indicators at the same times.

Zhou et al. found that the combination of guanidine ion and ferrocyanide ion can form a new thermosensitive crystal material, potassium ferrocyanide guanidine hydrate ($K_2(C(NH_2)_3)_2Fe(CN)_6{\cdot}6H_2O$) [7]. The material has low lattice energy and high dissolution entropy and shows excellent solubility temperature sensitivity. Due to the introduction of the heat-sensitive crystal material, a stable ferrocyanide ion concentration gradient can be formed at the hot and cold ends of the device in the presence of a temperature difference. For example, under the condition of 50 °C temperature difference, the concentration difference of ferrocyanide ion can reach 47 times, and the corresponding Seebeck coefficient increases from $1.4\,mV\,K^{-1}$ of the benchmark system to $3.73\,mV\,K^{-1}$. In addition, because there are a large number of undissolved crystal precipitates at the hot end of the device, the solution convection can be greatly inhibited and the effective thermal conductivity can be greatly reduced.

The preparation technology of low-temperature Bi_2Te_3-based refrigeration devices is relatively mature, and copper or aluminum is mainly used as electrode materials. Generally, tin soldering is used to realize the soldering of electrodes and Bi_2Te_3-based materials. Due to the limitation of the melting point of solder, it is difficult for such devices to be used above 200 °C. Electrode materials such as Al can be directly prepared on Bi_2Te_3-based thermoelectric materials by spraying technology. This device is suitable for thermoelectric power generation with heat source temperature below 300 °C because it does not use solder and has higher working temperature.

When applied to medium and high-temperature power generation devices, the selection and connection technology of high-temperature electrode materials is more difficult due to the high temperature. SiGe-based devices mainly use C, W, and Mo–Si as electrode materials [8], and the connection methods include spring pressure contact, hot isostatic pressing, hot pressing sintering, or spark plasma sintering. PbTe-based devices are mainly connected to Fe or Ni electrodes by hot-pressing sintering or spark plasma sintering, but the diffusion of Ni and Fe during the preparation process usually leads to the change of thermoelectric properties of PbTe-based compounds. Oxide thermoelectric power generation devices mainly use silver paste to directly connect silver electrodes, but silver paste is easy to volatilize at high temperature, which leads to device failure [9]. The electrode materials and connection technology of new CoSb-based filled kellogite devices have been widely studied. In the early days, NASA-JPL proposed the spring pressure contact method to prepare CoSb-based thermoelectric power generation devices, but the interface resistance and interface thermal resistance are large [10], which affects the improvement of device efficiency. Fan et al. rigidly connected Mo electrode and $CoSb_3$ material by using Ti transition layer and a two-step SPS method [11]. However, due to the large difference in thermal expansion coefficient between Mo and $CoSb_3$, the residual stress at the interface is large and cracks are easy to occur, which affects the bonding strength and device reliability. In this technique, the Mo–Cu thermal expansion coefficient can be adjusted to match the CoSb-base filled skutterudite by alloying Mo–Cu and optimizing the alloy composition, the residual stress at the interface is greatly reduced, and the interface strength and device reliability are improved [12, 13].

5.3.2 Device Construction and Functionalization of Thermosensitive Thermocells

The thermosensitive thermocell is generally formed by connecting an n-type thermoelectric material (n-type thermoelectric arm) and a p-type thermoelectric material (p-type thermoelectric arm) with a metal electrode (Figure 5.8). Because the output voltage of a single thermoelectric element is very low, it is usually necessary to connect many thermoelectric elements in electrical series and thermal parallel to form a practical thermoelectric device or assembly.

5.3.2.1 The Basic Structure of Thermosensitive Thermocells

π-shaped slab structure is the most typical and widely used device structure. In the π-shaped structure, the thermoelectric element composed of cuboid or cylindrical n-type and p-type thermoelectric arms is integrated into two electrically insulated and thermally conductive ceramic plates in the form of electrical series and thermal parallel, and the heat flow is transmitted along the direction perpendicular to the ceramic plates, so that the π-shaped structure is suitable for the working environment of a plate-shaped heat source (Figure 5.9). In the element flat-plate structure device, the heat flow density in the thermoelectric material is uniform, the unidirectional heat flow is easy to realize, and the thermoelectric performance of the material is easy to maximize through the optimal design of the device structure to realize high conversion efficiency. However, the cold and hot surfaces of the π-shaped planar device are usually in a constrained state in the working state, and the difference

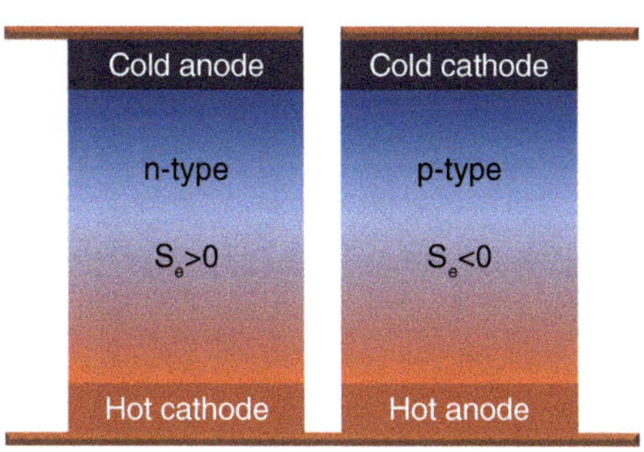

Figure 5.8 Schematic of integrated n-type and p-type thermocells.

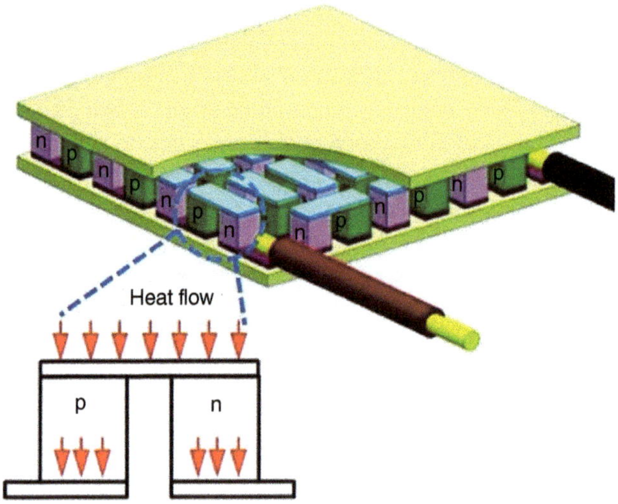

Figure 5.9 Structure diagram of π-shaped structure thermoelectric device.

in thermal expansion of different materials caused by the longitudinal temperature difference is easy to cause large thermal stress, which affects the reliability of the structure.

In addition to the traditional π-shaped structure, according to the requirements of different application environments and special service performance, a variety of non-element structures have been proposed, among which the most representative are annular structure and Y-shaped structure. In 2002, Weinberg et al. proposed that the thermoelectric device could be designed as a ring-shaped structure [14], as shown in Figure 5.10.

N-type and P-type ring thermoelectric materials (thermoelectric arms) are alternately arranged coaxially along the columnar heat source, and a layer of annular

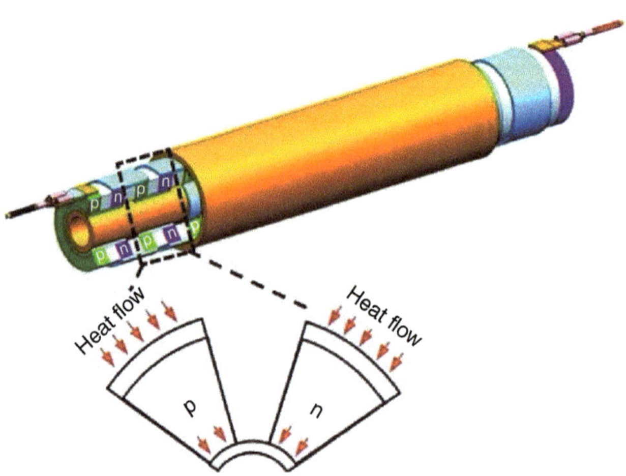

Figure 5.10 Structure diagram of ring-shaped structure thermoelectric device.

insulating material is placed between each annular thermoelectric arm to realize the electrical insulation between the adjacent thermoelectric arms. In addition, metal ring electrodes are arranged on the outer wall and the inner wall of the thermoelectric arm, and the outer metal ring and the inner metal ring are alternately arranged in sequence along the axial direction to realize the connection of adjacent thermoelectric arms and the electrical series connection of all p–n thermoelectric elements. The thermoelectric device with the annular structure is suitable for a non-flat heat source or a fluid medium cold and heat source, particularly the heat flow is transmitted along the radial direction of a cylindrical heat source, and the limitation that the traditional thermoelectric device with the n-shaped structure can only be applied to a flat heat source environment is favorably overcome. In addition, compared with a flat plate structure, it is easier to realize the integrated design and integration of the device and the heat collection/dissipation component, and it is easier to realize efficient heat transfer and low-stress structural design. However, the radial transmission characteristics of heat flow and current make it difficult to optimize the design of the temperature field and electric field. At the same time, the welding between special-shaped thermoelectric materials and metal electrodes and the integration technology of devices are more difficult than flat devices, and the manufacturing cost is also higher.

Thermoelectric power generation devices generally work in the environment of large temperature difference and cold and hot side constraints, so the thermal stress in the devices is inevitable, especially since most thermoelectric materials are brittle materials, and stress damage has become one of the outstanding problems in the engineering application of thermoelectric power generation devices. Stress-relieved or stress-free structures have become the most important goal in device design. Figure 5.11 shows a thermoelectric device with a Y-shaped structure. Cuboid or columnar n-type and p-type thermoelectric arms are alternately sandwiched between electrode connecting plates in a "sandwich" structure. The electrode

Figure 5.11 Structure diagram of *Y*-shaped structure thermoelectric device.

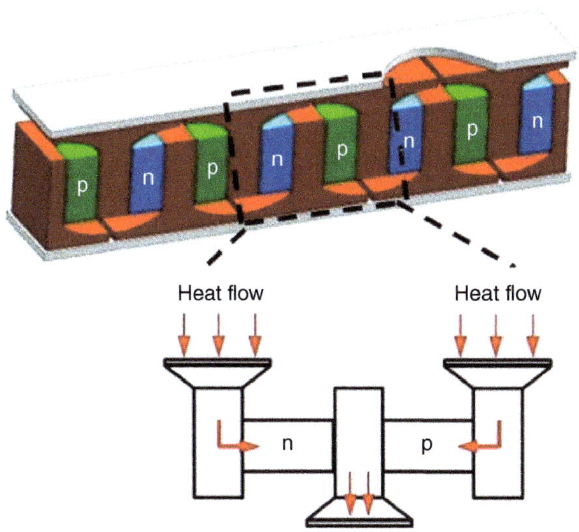

connecting plates not only provide conductive paths for adjacent p-type and n-type thermoelectric arms but also serve as heat collection and heat transfer members to form a "Y-shaped" heat transmission channel together with p-type and n-type thermoelectric materials. This Y-shaped structure makes the current flow parallel to the cold and heat sources, that is, the direction of heat flow is perpendicular to the direction of current flow. The Y-shaped structure separates the thermoelectric material from the stress field, the device mounting stress is applied on the high and low-temperature electrodes and the insulating part and is not directly transferred to the thermoelectric material or is transferred to the thermoelectric material after being relieved, the thermoelectric materials are kept in a relatively independent or incomplete constraint state, and the constraint degree of the thermal expansion deformation on the cold and hot surfaces is reduced, thereby being beneficial to the stress relieving structural design of the thermoelectric materials and structural parts. The transverse serial thermal expansion of the n-type and p-type thermoelectric arms avoids the stress concentration caused by the difference of expansion coefficients and provides greater flexibility for the height and area design of the thermoelectric arms. In addition, the structure enables each thermoelectric element to be independently optimized, and each thermoelectric arm can have different heights and interface structures, which is beneficial to the manufacture of segmented structure devices. However, the heat flow and current density in the thermoelectric material in the Y-shaped device are not uniform, which will affect the maximum performance of the thermoelectric material and reduce the efficiency of the device to a certain extent. In addition, the electrode connecting plate which almost runs through the device and connects the high and low temperature ends also causes large heat leakage, which also makes it difficult to improve the conversion efficiency of the device.

5.3.2.2 The Interface Structure of Thermosensitive Thermocells

For the thermosensitive thermocells used for a long time, the element interdiffusion or chemical reaction is easy to occur at the interface between the high-temperature end electrode and the thermoelectric material, which leads to the change of the interface composition and structure, and produces additional interface resistance and thermal resistance, thus causing the performance degradation of the device, and even leading to the failure of the device. In addition, the matching of high electrical conductivity, high thermal conductivity, and thermal expansion coefficient with thermoelectric materials is the first consideration in the selection principle of most metal electrode materials, but these electrode materials are often difficult to directly form a good combination with thermoelectric materials, that is, it is difficult to achieve practical application requirements.

In order to effectively solve the problems of good connection (high strength, low interfacial resistance, and thermal resistance) and high-temperature interfacial stability between the electrode material and the thermoelectric material, an appropriate transition layer can be introduced between the thermoelectric material and the electrode. The main function of this transition layer is to improve the wettability between the material and the solder and to reduce the interfacial resistance and thermal resistance for the soldering process of low-temperature devices. For example, nickel plating on the soldering surface of bismuth sulfide material is the most commonly used and effective method for bismuth sulfide devices, but for high-temperature devices, the selection and design of the transition layer are more complex, and three functions need to be considered: playing the role of "activation layer" in the electrode preparation process, promoting the combination of thermoelectric materials and electrodes, and improving the bonding strength. A stable interface layer is formed, the interfacial layer is used for blocking the mutual diffusion and chemical reaction between the thermoelectric material and the electrode at high temperature, and the resistance and thermal resistance of the interfacial barrier layer itself must be low enough, that is, the interfacial barrier layer must be a good conductor at high temperature.

5.3.3 Performance Test Method of Thermosensitive Thermocells

5.3.3.1 Evaluation and Measurement of Conversion Efficiency and Output Power

The maximum energy conversion efficiency and the maximum output power are the main indexes to evaluate the performance of the device. The schematic diagram of the thermoelectric device conversion efficiency evaluation device was shown in Figure 5.12. Generally, a temperature difference is established at both ends of the device (high temperature: T_h, cold temperature: T_c), the load resistance is changed under the temperature difference, the corresponding device output voltage (V_{out}) under different loads and the voltage drop (V_s) on the standard resistance (R_s) are measured, and the device output current (I_{out}) can be obtained by the following formula:

$$I_{out} = V_s/R_s \tag{5.13}$$

As shown in Figure 5.13, I_{out}–V_{out} linear relationship and I_{out}–P_{out} curve can be obtained by changing the external load resistance value. When the R_{in} of a single device is relatively small, the output current I_{out} and the output voltage P_{out} can

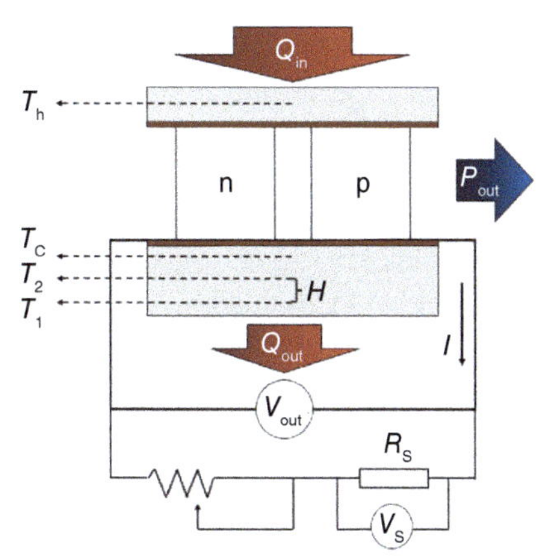

Figure 5.12 Schematic diagram of evaluation device for thermoelectric power generation devices.

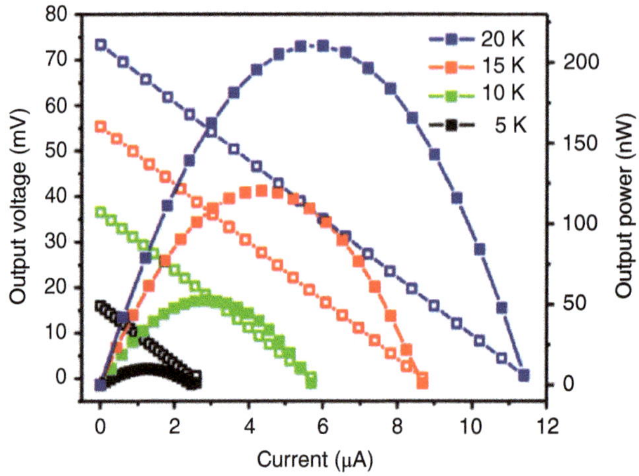

Figure 5.13 Schematic diagram of device performance measurement parameters. Source: Jung et al. [15]/American Chemical Society.

be linearly fitted to obtain the internal resistance R_{in} and the open circuit voltage V_{oc} of the thermoelectric device under the temperature difference. According to the I_{out}–V_{out} linear relationship and the I_{out}–P_{out} curve, the maximum output power P_{max} (corresponding to the output power when the total resistance of the external load is equal to the internal resistance) of the device under this temperature difference condition can be obtained.

The heat flow Q_{out} discharged from the cold end is measured by a copper block (heat flow meter) with known thermal conductivity, that is, the temperatures T_1 and T_2 at the high and low temperature measuring points of the heat flow meter and the geometric dimensions of the heat flow meter are accurately measured, then:

$$Q_{out} = \kappa \times \frac{T_1 - T_2}{H} \cdot (L \times W) \tag{5.14}$$

where κ is the thermal conductivity of the heat flow meter, L and W are the two side lengths of the section of the heat flow meter, and H is the vertical distance between the two temperature measuring points T and T on the heat flux meter. According to the definition, the conversion efficiency of the device can be calculated:

$$\eta = \frac{P_{out}}{Q_{in}} \times 100\% = \frac{P_{out}}{P_{out} + Q_{out}} \times 100\% \tag{5.15}$$

For a given temperature difference, the $I_{out} - \eta$ curve and the maximum conversion efficiency η_{max} at the corresponding temperature difference can be obtained.

In the device evaluation, it is necessary to maintain a stable large temperature difference environment for a long time, and there are many measurement parameters, which are easy to produce large test errors.

The main sources of test errors are as follows:

(i) It is difficult to accurately control the temperature of cold and hot ends. Usually, it is necessary to measure a series of output voltages by changing the load under the same temperature difference condition. With the increase of the test current, the heat absorbed by the hot side and the heat discharged by the cold side due to the Peltier effect will increase, which will usually lead to a decrease in the temperature of the hot side and an increase in the temperature of the cold side, affecting the accurate measurement of the output voltage and current of the device. Therefore, it is very important to accurately control the temperature of cold and hot ends.

(ii) Heat radiation and convection heat transfer are difficult to avoid. There is a large temperature difference between the heat source and the cold source (especially when measuring at high temperature), and the radiation heat transfer and convection heat transfer increase. The heat flow measured by the heat flow meter at the cold end is not entirely the heat conducted through the device, which directly affects the accurate acquisition of the conversion efficiency. In general, applying thermal insulation around the device and maintaining a high degree of vacuum helps to improve the measurement accuracy. In addition, the accurate measurement of the temperature at the hot end of the device and the calibration of the thermal conductivity of the heat flow meter will have a significant impact on the measurement results of the device's performance.

5.3.3.2 Measurement of Thermoelectric Properties of Materials by Harman Method

The *ZT* value of the thermoelectric material can also be obtained by using the evaluation test method of the device performance. The basic method to obtain the dimensionless thermoelectric Figure of merit is to measure the electrical conductivity, Seebeck coefficient, and thermal conductivity, respectively, and then calculate the thermoelectric figure of merit of the measured material according to the definition of *ZT* value. However, this evaluation method is complex and time-consuming because it involves a large number of tests. The *ZT* value of the thermoelectric material itself can also be obtained at one time by testing the performance of the final thermoelectric component, which is very simple and fast. For example, the thermoelectric figure of merit can be estimated from the maximum temperature difference of the device. The relationship between the maximum temperature difference T of the device and the thermoelectric figure of merit under ideal conditions can be expressed as:

$$Z = \frac{2\Delta T_{\text{max}}}{T_{\text{c}}^2} \tag{5.16}$$

where T is the cold junction temperature of the thermoelectric device. The maximum temperature difference T_{m} has an approximate parabolic relationship with the current applied to the device, so the maximum temperature difference can be easily determined by measuring the relationship between the heat of the device and the current, and then the thermoelectric figure of merit can be inversely derived by using the above formula. Obviously, this method is not only simple but also convenient. However, it must also be pointed out that this method is only an approximate estimation of the thermoelectric figure of merit.

Seebeck coefficient, electrical conductivity, and thermal conductivity are the properties of materials. We know that the basic principle of thermoelectric power generation devices is the Seebeck effect, which generates electricity through thermoelectric potential when there is a temperature difference between the two ends. At this time, the thermoelectric material is used as a power supply. Under the same temperature difference, the larger S means the larger electromotive force, while the larger conductivity means the smaller internal resistance of the power supply.

The thermal conductivity can be divided into two parts: One part of the heat from the heat source will be converted into electricity and another part of it will be transported to the cold end. The greater the thermal conductivity, the more heat will be "wasted." To use the Seebeck effect, it is necessary to establish a temperature difference between the two ends of the material. If the thermal conductivity of the material is greater, it will be more difficult to establish a temperature difference.

5.4 Summary and Perspective

Since the late 1990s, the research of thermoelectric materials has made rapid progress. The discovery of a variety of new material systems and the proposal of a new thermoelectric transport performance control theory have led to the continuous increase of thermoelectric figure of merit. On the other hand, although the application of thermoelectric conversion technology has received extensive attention, the large-scale industrial application of thermoelectric technology, especially the technology of thermosensitive thermocell, is progressing slowly. The R&D chain of thermoelectric conversion technology is long, and the development of back-end technology of devices and systems is difficult, which creates a long R&D path from materials to applications. First of all, as the core and bottleneck of the large-scale application of the technology of thermosensitive thermocell, the development of device technology lags behind the rapid development of thermoelectric material science. The high performance of devices is faced with more complex scientific and technical problems that need to be solved. For example, the optimal design method of the device topology structure needs to be further improved, and the electrode preparation technology with high reliability and low loss needs to be breakthrough. Although the technology of thermosensitive thermocell has the characteristics of maintenance-free and long life because it has no moving parts, the service attenuation and failure of devices in harsh environment are inevitable, and the prediction system of device reliability and life is not perfect, which directly affects the popularization and application. Second, as an alternative to traditional energy conversion technology, the cost advantage of the technology of thermosensitive thermocell is not significant compared with existing technologies. The establishment of low-cost batch manufacturing technology from materials to devices and systems also requires the innovation of source technology, especially the batch manufacturing and low-cost technology of devices, which is the key to determine whether thermosensitive thermocells technology can move from "laboratory" or "small-scale application to large-scale industrialization." It is also the key to determine whether the research field of thermoelectric materials can continue to prosper. Finally, the advantages and disadvantages of thermosensitive thermocells technology are obvious, and the application fields of thermoelectric conversion technology characterized by decentralization, low power, and low energy density are determined by its own characteristics. The development of new technology in the application of thermosensitive thermocell needs to return to the origin of developing its strengths and avoiding its weaknesses. It is necessary to comprehensively reflect the overall factors such as irreplaceability, technological advancement, and economic or social benefits of the technology of thermosensitive thermocell. Aiming at the specialty application field and developing the irreplaceable the technology of thermosensitive thermocell is the development direction of this field.

References

1 Rahimi, M., Straub, A.P., Zhang, F. et al. (2018). *Energy Environ. Sci.* 11: 276.
2 He, J. and Tritt, T.M. (2017). *Science* 357: eaak9997.
3 Zhu, J., Yao, M., Huang, S. et al. (2020). *Angew. Chem. Int. Ed.* 59: 16480.
4 Wu, K. (2011). *Mach. Electr. Appliances* 7: 43.
5 Jianming, C. (2013). *Inner Mongolia Sci. Technol. Economy* 19: 76.
6 Liu, X. (2011). *Metrol. Test Technol. Verifi.* 21: 86.
7 Yu, B., Duan, J., Cong, H. et al. (2020). *Science* 370: 342.
8 Lin, J.-S. and Miyamoto, Y. (2000). *J. Mater. Res.* 15: 647.
9 Lim, C.-H., Choi, S.-M., Seo, W.-S., and Park, H.-H. (2012). *J. Electron. Mater.* 41: 1247.
10 El-Genk, M.S., Saber, H.H., and Caillat, T. (2003). *Energy Convers. Manage.* 44: 1755.
11 Fan, J., Chen, L., Bai, S., and Shi, X. (2004). *Mater. Lett.* 58: 3876.
12 Zhao, D., Li, X., He, L. et al. (2009). *J. Alloys Compd.* 477: 425.
13 Zhang, Q.H., Huang, X.Y., Bai, S.Q. et al. (2016). *Adv. Eng. Mater.* 18: 194.
14 Weinberg, F.J., Rowe, D.M., and Min, G. (2002). *J. Phys. D: Appl. Phys.* 35: L61.
15 Jung, J., Suh, E.H., Jeong, Y.J. et al. (2019). *ACS Appl. Mater. Interfaces* 11: 47330.

6

Wearable Power Generation via Thermoelectrochemical Devices

6.1 Introduction

During the surge of development, global energy demand reached a peak point. However, it continued to increase by more than 30% between the twentieth century and the middle of the twentieth century, which is equivalent to adding two more countries with energy demands similar to India at the time. As the global population increases from seven billion people today to over nine billion in the future, and the global economy expands at a fast rate of 3.4%, the world's energy supplies face a significant challenge.

At present, the burning of fossil fuels can provide the majority of our energy for life, which leads to an increase in the cost of fossil energy, and at the same time, the global environmental impact of its burning has led most scientists to take a keen interest in renewable and sustainable energy sources, such as efficient, nonpolluting renewable energy sources that are being developed and explored (e.g. solar, wind, wave, tidal, acoustic, etc.), which are considered to be alternative solutions to future energy problems. Waste heat cogeneration is getting noticed and is consistently regarded as a contender for green and clean energy among the various energy technologies that have emerged in recent years.

In the field of mobile electronics, batteries are one of the most common power sources and have a wealth of experience and industrial scope. However, these batteries often fail within a limited period of time due to short-circuiting, thermal runaway, and low charge. As a result, they often require manual replacement with new batteries. For example, wearable medical instruments that provide health tracking for the aged and communicate physiological data of the user frequently need energy storage devices to provide a continuous power source. These devices cannot be simply changed with batteries or other tools because they work inside the patient's body for many years and require the use of wireless sensors [1–3].

However wearable electronics have never been able to use standard renewable energy sources because of the complexity, expense, and bulk of their design. Therefore, one potential answer is to power these wearable devices using thermoelectric generators, which can produce energy and collect waste heat from the human body. Through physiological processes, the human body serves as a significant source of

Low-Grade Heat Harvesting: Materials, Devices, and Technologies, First Edition. Xiaogang Zhang.

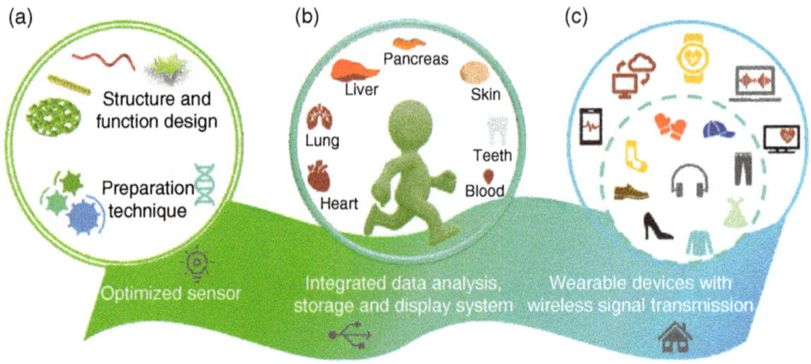

Figure 6.1 Schematic diagram of the future development of wearable intelligent monitoring devices. Source: Dai et al. [4]/John Wiley & Sons/CC BY 4.0. (a) material structure and function design depending on preparation technique, and (b, c) wearable monitoring device integration with wireless data analysis, storage and display system for healthcare.

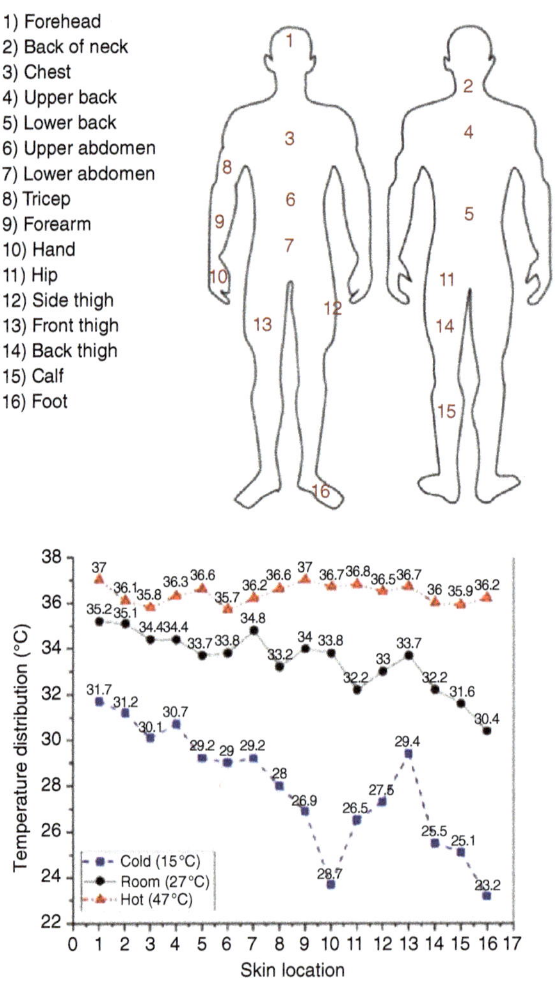

Figure 6.2 Reported human skin temperatures for different points on the body at varying ambient temperatures. Source: Adapted from Suarez et al. [8].

thermal energy and continuously produces heat. The resulting handheld electrical gadgets have a maximum power output of a few watts (Figure 6.1) [5–7].

For example, as shown in Figure 6.2, when the human body is at room temperature between 15 and 47 °C, capturing the heat emitted by the body using a thermoelectric generators (TEGs) with a conversion efficiency of approximately 1% generates approximately 1–5.25 W of power. This is sufficient to power most wearable sensors. However, it requires covering the entire body with the TEG, which is not very feasible for practical applications. Thank goodness, with the development of miniaturization and lightweighting of electronic devices, the power requirements of sensors and micro-signal processors have dropped from the milli-watt to micro-watt (W) level. As a result, thermoelectric devices are quickly replacing other platforms for supplying electricity to peripheral technology [8–11].

6.2 Thermoelectrochemical Devices Requirements for Wearables

At the equipment level, TEGs are designed to maximize their power output, so several factors in the design of their equipment must be carefully considered.

6.2.1 Filler Material

The filler material is usually a fixed layer during device manufacture or an intentional addition material. It is attached to the hot and cold sides of the device and is generally used to provide structural and mechanical support for pillars and top interconnects.

The lateral heat loss of the filler material reduces the temperature difference and reduces the output power of the TEG. The ratio of the flat area occupied by the thermoelectric material to the total surface area of the module is called the filler factor. As an example, for a device where the total thermal conductivity of the thermoelectric material and the filler are equal; this would result in at least half of the heat loss in the device being caused by the filler material. Therefore, a filler material with a high and low thermal conductivity is an important means of effectively increasing the power of the device. After a selection of various materials, air, as a filler material with low thermal conductivity, only 1% less power than the vacuum and 60% higher power output than SiO_2, does not require sealing during working and exhibits a wide range of applications. Moreover, various aerogels, such as silica, have an even lower thermal conductivity than air and its filling can effectively increase the mechanical strength of the device and maintain structural integrity while adding minimal mass to the module (Figure 6.3).

6.2.2 Thermal Load Matching

Device performance is mainly determined by the thermal resistance of the TEG, which under optimal conditions must be as high as the parasitic thermal resistance

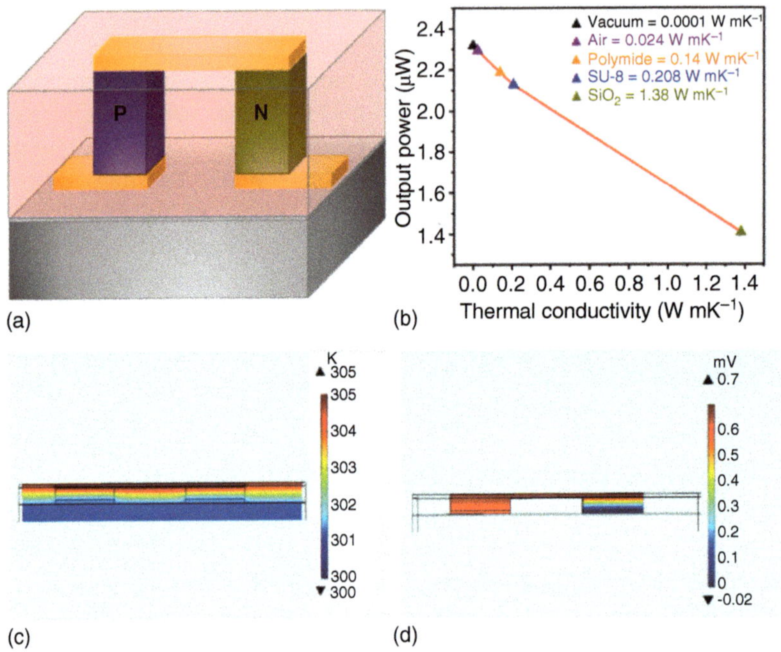

Figure 6.3 (a) Schematic of a single-leg pair TEG with SiO_2 as filler material; (b) power output of single-leg pair TEG with different filler materials; (c) temperature; and (d) potential distribution of single-leg pair TEG with SiO_2 as filler material. Source: Tanwar et al. [10]/MDPI/CC by 4.0.

of the TEG in series. The thermal resistance of a TEG can be increased in several ways: by reducing the number of *p* and *n* legs, by increasing the aspect ratio of the legs, or by reducing the thermal conductivity of the legs. First, skin-to-TEG resistance when placed on the skin can be seen as the two main resistances between skin temperature and core body temperature. The first of these resistances is the skin resistance, which is a complex function of a number of human physiological indicators formulated together including gender, age, and weight. The second resistance is the thermal contact resistance at the TEG–skin interface, and as the skin is not a smooth plane, the contact resistance with the TEG can have a large impact on its performance. The contact resistance at the skin/TEG interface is high pressure dependent as the skin is not a smooth plane and the contact resistance with the TEG has a large impact on its performance. However, under pressure, the skin is deformed due to its softness. The third resistance is the thermal resistance of the TEG to the ambient interface, which depends on the air convection. If a heat sink is not used on top of the TEG, the general TEG–air interface resistance is therefore usually the main thermal resistance.

In order to achieve maximum power output, it is essential to match the thermal resistance between the TEG and the thermal generator. From the point of view of thermal circuit construction, where the thermal generator consists mainly of the body and air and the thermal load consists of the TEG and its corresponding parasitic resistance, the optimum thermal resistance of the TEG can be defined as:

$$R_{\text{TEG,Optimal}} = \frac{R_F(R_{\text{skin}} + R_{\text{air}})}{2(R_{\text{skin}+R_{\text{air}}}) + R_F} \tag{6.1}$$

where

R_F: is the parasitic thermal resistance
R_{skin}: is the thermal resistance between the body core and the surface of the skin
R_{air}: is the thermal resistance of the ambient air.

6.2.3 Lateral Heat Flow and Substrates Effecting

For TEG devices, the substrate is an important component that collects and releases the heat flowing laterally. This substrate on the hot side to collect the heat from the skin and then transfer it to the legs and also on the cold side to release the heat into the heat sink for convection through the substrate. It is assumed that a material with a thermal conductivity of $\kappa = 200\,\text{W m}^{-1}\,\text{K}^{-1}$, a value close to the thermal conductivity of aluminum nitride ($\kappa \approx 185\,\text{W m}^{-1}\,\text{K}^{-1}$ at 300 K) and aluminum ($\kappa \approx 205\,\text{W m}^{-1}\,\text{K}^{-1}$ at 300 K), can be used as a substrate on a flexible TEG (Figure 6.4).

In Figure 6.5, the relationship of power vs. fill factor is shown for different substrate thicknesses of 0.5, 5, 50, and 500 μm, where air is the fill material. Under these conditions, a flexible TEG can be constructed for the device. The results in the graph show that the effect of substrate thickness on performance is minimal at thicknesses above 5 μm. This means that a flexible thermoelectric material can be constructed without sacrificing overall performance. Sufficient thermal dispersion can be achieved by using thinner substrates or metal diffusers.

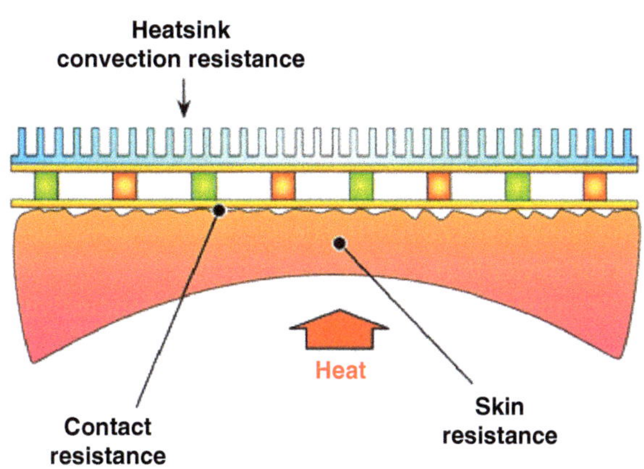

Figure 6.4 Cartoon illustrating the three parasitic resistance that impacts TEG performance. Source: Suarez et al. [8]/Royal Society of Chemistry.

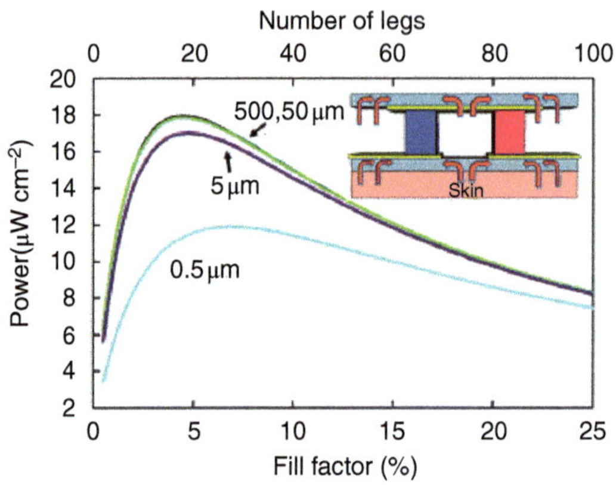

Figure 6.5 The output power for different substrate thicknesses. Source: Suarez et al. [8]/Royal Society of Chemistry.

6.3 Materials Toward Wearable Devices

6.3.1 Inorganic Material

In general, inorganic materials are always ground to a powder or melted/heated to form a powder when used as thermoelectric components, which facilitates the acquisition of ingot legs and the subsequent consolidation of the prepared material into ingots by cold pressing or hot pressing. The typical organic materials conclude the Bi_2Te_3 or $Bi_{0.3}Sb_{1.7}Te_3$ composites for p-type and the Bi_2Te_3 or $Bi_2Te_{2.7}Se_{0.3}$ for n-type.

Suarez et al. propose a strategy of using polydimethylsiloxane (PDMS) as a flexible substrate and applying it to a flexible thermoelectric device in which the thermocouple consists of p-type ($Bi_{0.5}Sb_{1.5}Te_3$) and n-type ($Bi_2Se_{0.3}Te_2$) and with eutectic gallium–indium (EGaIn) to electrically connect the ingot-shaped legs together. This EGaIn not only offers extremely low interconnection resistance but also self-healing stretch, both of which are essential for flexible TE modules. When the device is worn on the wrist at 24 °C ambient temperature (~37 °C) using a TEG consisting of 32 pairs of legs, it can deliver 1 µW cm^{-2} of power. Due to the lower thermal conductivity of PDMS compared to air, this results in lower output voltages when using PDMS to package the device legs compared to conventional TEGs. Meanwhile, the resistance does not change after 1000 bending cycles with the device, indicating the high mechanical strength of the device. This demonstrates the high mechanical strength of the device [12].

In order to further reduce the thermal conductivity of the PDMS substrate, Jung et al. [13] proposed the use of a sponge-like PDMS as a filler, which was prepared by adding sugar to a stainless steel mold and then using a hydraulic press to produce a block-shaped sugar template (approximately 60 mm in diameter and 1.15 mm in thickness), the pores of which would determine the filler content. The sugar block template was then dipped into the PDMS matrix and curing agent, and after curing was complete the sugar template was placed in water to obtain spongy PDMS. The prepared PDMS sponge was inserted into a Bi_2Te_3-based thermocouple measuring

1.2 mm × 1.2 mm × 2.6 mm and the thermocouples were connected together using copper foil as the electrode. At the same time, the relatively thin packing (~1.2 mm thick) allows for a reduction in the tensile stresses acting on the electrodes and legs, so that the device can be bent repeatedly for 2000 cycles without an increase in ΔR, demonstrating its mechanical flexibility. The thermal conductivity of the porous PDMS was 0.08 W m^{-1} K^{-1} which is much lower than the dense PDMS (0.08 W m^{-1} K^{-1}), thus TEG device with porous PDMS filler can effectively improve the ΔT, output voltage, ZT, and power density and produce early 130 μW cm^{-2} at a temperature difference of 8 °C. At the same time, the thermal resistance model developed by the authors illustrates that thermal resistance is a strong function of filler thickness and thermal conductivity. Surprisingly, in the final demonstration experiment, the authors used 8 TEGs connected in series to make 32 LEDs glow for over 30 minutes.

To reduce the effect of conventional polymers on the legs, Hong constructed rigid TE columns using $Bi_{0.5}Sb_{1.5}Te_3$ (p-type) and $Bi_2Te_{2.7}Se_{0.3}$ (n-type), which were electrically connected together by copper tape as electrodes. The authors then sandwiched the TE columns between a commercially available flexible polymer called Ecoflex. Specifically, only the top and bottom sections of the TE column were embedded in the Ecoflex plate, a design that allows a large part of the leg to be exposed to air. This double elastomeric layer design allows the TE column to be flexible enough for high aspect ratios and to increase the heat dissipation effect. The device demonstrates long-term (>8 hours) and large (>10 °C) active cooling effects and exhibits an extremely high output power of 4.5–25.1 μW cm^{-2}. At the same time, the authors integrated the device into a TED garment, which kept the skin at a thermally comfortable temperature of 32 °C at widely varying ambient temperatures (22°–36 °C) [14].

Wang focused on the use of flexible printed circuit boards (FPCBs) with special holes as substrates to replace rigid electrodes, and the authors used 52 pairs of p-($Bi_{0.5}Sb_{1.5}Te_3$) and n-($Bi_2Se_{0.5}Te_{2.5}$) thermoelectric legs arranged in 4 columns and 13 rows, connected by soldering to the bottom FPCB and the top by copper tape, with overall dimensions of 26.5 mm wide and 43.5 mm long and 26.5 mm wide and 43.5 mm long. A polyimide (PI) film is also used to insulate the TEG from the skin surface of the wrist. In the design of the FPCB, additional holes are added to the structure to further increase its bendability. The use of PDMS as a soft packaging material reduces heat loss and improves performance. Based on the above design results, the TEG device exhibits a low internal resistance (1.8 Ω) and an open circuit voltage of 37.2 mV at $\Delta T = 50$ K. This corresponds to an output power of 192.6 μW and 16.7 μW cm^{-2}. The TEG device can generate 6.6 mV at $\Delta T = 5.8$ K when worn on the wrist and can successfully power a miniature accelerometer to detect body movements under different motion conditions [15].

Liu has developed a novel assembly strategy where the legs are not encapsulated in a whole PDMS block, but rather each of them is encapsulated in a separate PDMS block, whereby the FECB substrate, 18 pairs of Bi_2Te_3 (2 mm × 2 mm × 1.5 mm) legs, cold ends are connected with silver paste to build the TEG device and a separate PDMS layer is poured on the side of each leg to achieve a separate packaging. At the

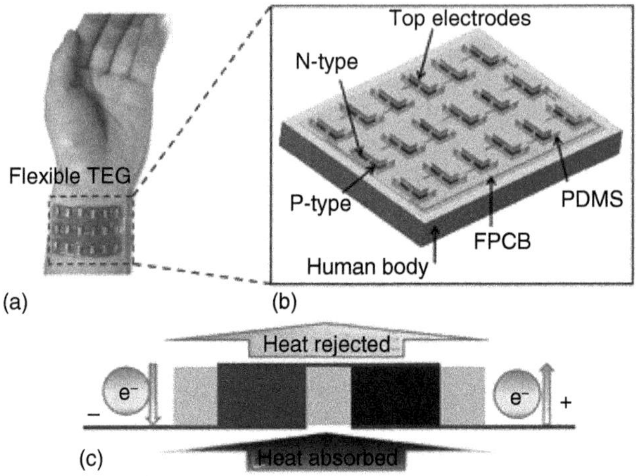

Figure 6.6 Schematic diagram of the designed wearable TE device. Source: Liu et al. [16]/Springer Nature.

same time, PI films can be used to insulate the device from the skin. Devices using this strategy become more flexible and can produce 8.3 μW output power (approx. 0.37 mA and 22.4 mV) at 11 K temperature difference [16] (Figure 6.6).

In order to further reduce the effect of polymers on thermal conductivity, the authors did not use PDMS as a filler material but exposed it directly to air. As the bismuth telluride material has the highest thermoelectric efficiency near 300 K compared to other thermoelectric materials, 47 pairs of p-type ($Bi_{0.5}Sb_{1.5}Te_3$) and n-type ($Bi_2Te_{2.8}Se_{0.2}$) legs were soldered to the FPCB from the bottom and the tops were soldered together using copper strips. In addition, the bracelet can be connected to a variety of smart devices to continuously monitor the body's temperature, humidity, and human activity while relying solely on body temperature, humidity, and human activity under body temperature alone. In addition, the authors demonstrate a correlation between the output voltage and the height of the leg as well as the fill factor [17].

Eom developed a rigid inorganic bulk material-based flexible thermoelectric system (FTES) that resembles a watch bracelet, and they contain 10 thermoelectric system units, each with four TE elements in a 2 × 2 array with a heat sink on top and p- and n-type legs made of $Bi_{0.5}Sb_{1.5}Te_3$ and $Bi_2Te_{2.7}Se_{0.3}$ materials, with dimensions of 3 mm × 5 mm × 5 mm, which are connected together by shafts that penetrate the elements and links. In total, the FTES has 40 TE elements and measures 156 mm × 16 mm, making it very adaptable to human wear. The device exhibits excellent flexibility, preforming in bending tests with comparable performance to conventional rigid modules, but theoretical analysis shows that FTES has a high resistance due to the gap between the device and the bending heat source, which increases the contact resistance. The authors suggest that the use of thick copper wire electrodes is the main cause of this phenomenon and that the use of a thinner flexible unit would eliminate this effect. When the FTES was worn on the human body, the maximum output power obtained was 2, 2.9, and 4.7 $\mu W\,cm^{-2}$ for sitting, walking, and running conditions, respectively, at a ΔT of 20 °C. The device was able to generate approximately 80 µW of power when running at a slow speed even with high thermal contact resistance [18].

Siddique et al. have changed the process of preparing the leg, not by cold/hot pressing the TE powder, but by preparing a paste of TE and printing it into a highly heat-resistant fabric already prepared in polyester. This method has the advantages of simplicity, low cost, and low curing temperatures (i.e. between 100 and 200 °C). First, 24 holes are cut out of the preprepared fabric and the n-type $(0.98Bi,0.02Sb)_2(0.9Te,0.1Se)_3$ and p-type $(0.25Bi,0.75\ Sb)_2(0.95Te,0.05Se)_3$ TE materials are mixed with 18–20 wt% of Durabond-950 binder solution to obtain a paste. The paste-like material was obtained by mixing well. The prepared n- and p-type pastes were then used to fill the holes of the fabric substrate alternately and after the curing process, the ingot legs were joined together with silver wire to act as electrodes. Finally, the Kapton film is attached to one side of the FTEG which will be attached to the human skin. When the TEG was tested as a device and placed on the arm, the highest achievable output power was 1×10^{-5} µW cm^{-2} at a ΔT of 32.9 °C [19].

Varghese et al. used microwave-stimulated wet chemical synthesis to synthesize doped and functional $Bi_2Te_{2.8}Se_{0.2}$ nanocrystals, which exhibited platelet-like structures several tens of nanometer thick. The material was homogeneously mixed with a binder to form an ink and subjected to dry (58 wt% $Bi_2Se_{2.8}Se_{0.2}$ nanopowder, 39 wt% solvent, and 2 wt% binder), and five 10 mm × 2 mm × 0.01 mm n-type $Bi_2Te_{2.8}Se_{0.2}$ elements were printed onto flexible polyimide substrates with 4 mm spacing, using copper foil to connect the originals. These printed TE films showed an unprecedented peak ZT of 0.43 at 175 °C and no significant change in thermal resistance after 150 bending tests, and the thermoelectric element constructed using this film of n-type material exhibited a maximum output power of 500 µW cm^{-2} at a temperature difference of 20 °C [20] (Figure 6.7).

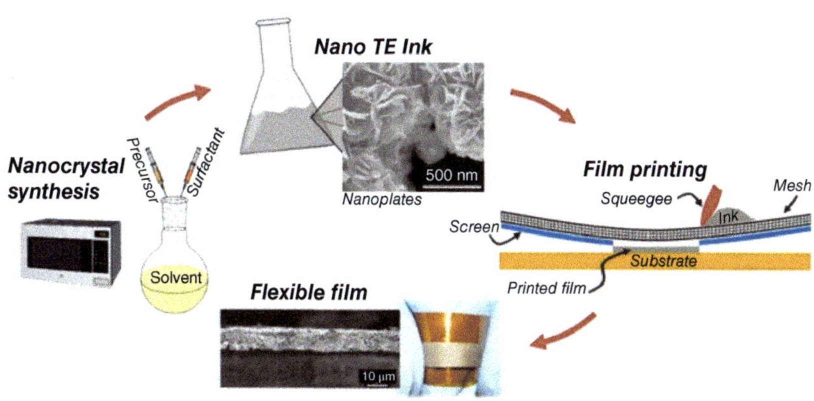

Figure 6.7 Schematic illustration of the overall fabrication process for the flexible thermoelectric films. Source: Varghese et al. [20]/Springer Nature/CC BY 4.0.

Guo et al. reported on a MoS_2 material used to replace conventional inorganic TE materials. In this work, the authors used in situ growth to achieve Au doping of MoS_2 and demonstrated that metal doping can effectively enhance the thermoelectric properties of the material. By TEM and SEM characterization, the Au nanoparticles were approximately 5–10 nm in size and randomly distributed on the 2D MoS_2 surface. The authors made thin films of Au-MoS_2 and cut them into 5 mm × 45 mm strip structures, which were subsequently passed through the holes of the prepared fabric and finally connected using copper wires. The TEG device achieves a power factor ($\sigma S2$) of 166.3 $\mu W\,m^{-1}\,K^{-2}$ at room temperature, an enhanced σ and Seebeck coefficient (S) due to Au doping, as well as a maximum output voltage of 2.5 mV at ΔT at 5 °C [21].

Chen demonstrated a technique for constructing TEGs using inkjet printing by inkjet printing bismuth telluride (Bi_2Te_3) and bismuth telluride antimony ($Bi_{0.5}Sb_{1.5}Te_3$) nanowires (NWs) onto polyimide and using them as n-type and p-type branched circuits. After printing, the authors used an annealing process to enhance the hot spot properties of the TE material. It is worth noting that the metal contacts in the TEG are connected by EGaIn liquid metal. Finally, the TEG device, constructed with five pairs of p-n branches, was able to deliver 45 mV, and the printed TEG provided a maximum output power of 14.1 nW at a temperature difference of 7 °C. The thermoelectric properties of the TEG devices did not degrade after several bending experiments (up to 50 times) [22].

Ren et al. constructed a high-performance TEG device with excellent stretching, self-healing, recycling, and reassembly capabilities. First, to prepare thermoelectric chips, four pairs of $Bi_{0.5}Sb_{1.5}Te_3$ and $Bi_2Te_{2.8}Se_{0.3}$ were alternately deposited onto polyimide films as n- and p-leg, respectively. After treatment with an argon atmosphere at 320 °C, the TE legs were joined together by Au–Ge electrodes. For modular assembly, a laser-cut polyimide substrate is used to create the slots and screen-print the patterned liquid metal wires. The modular thermoelectric chip is then inserted into the slot in the polyimide substrate and a small amount of polyimide solution is dripped over the connection to ensure a stable connection between the chip and the polyimide film. The device can produce a record open circuit voltage density of $1\,V\,cm^{-2}$ at a temperature difference of 95 K. This special structure allows the rigid and fragile TEG chip to be strain separated from the soft polyimide substrate during mechanical deformation, thereby it can accommodate large angles of bending. At the same time, the liquid metal wiring and bond exchange reactions in the polyimide network ensure that the devices heal themselves even in the event of fracture and generate sufficient covalent bonds at the interface to guarantee the mechanical strength of the connection. The stand-alone TE devices can be customized for different conditions in terms of form factor, structure, output voltage, and power [23].

To improve the integration of the device, Wen et al. printed $Bi_2Te_{2.7}Se_{0.3}$ (n-type) and Sb_2Te_3 (p-type) as the thermoelectric leg and thermoelectric electrode on a PI film to form a double chain structure, with the gap between the two chains filled with silk protein in the middle, with an overall size of 60 mm × 35 mm × 0.2 mm. It is noteworthy that the TE paste is printed in the form of a double chain printing, which ensures more capacitive effects for functional sensing with no change in power. The different adsorption behavior of the filamentous protein layer toward different states of water in air (water vapor, small droplets), the absorption and desorption of water molecules change the strength of their molecular motion, leading to a change in the dielectric constant. The TEG device can therefore not only achieve high performance in thermal differential power generation (maximum output power of 0.095 μW cm^{-2} at ΔT at 20 °C) but also detect changes in water and temperature in air. To further verify its feasibility, this TEG device was charged with 2200 μF capacitors by connecting them to a series-parallel switching circuit to achieve a 3.3 V output [24].

Karthikeyan et al. also applied the thermal evaporation technique to evaporate TE materials (p-type SnTe and n-type PbTe) in a vacuum and solidify them as thin films on a flexible polyimide substrate, which is interconnected with aluminum to form a thermoelectric generator structure. These deposited films exhibit a maximum ZT ~0.3 in p-type and ~0.2 in n-type at 550 K. The TEGs fabricated by this process provide an output power of 8.5 mW cm^{-2} and a current density of 34.7 mA cm^{-2} at a temperature difference of 120 °C. The ability of the device to sustain temperature differences was also collated using infrared images and power density observations [25] (Figure 6.8).

6.3.2 Organic Polymer Material

Although inorganic materials have many advantages and are well researched, they cannot withstand a wide range of dynamic movements when combined with flexible devices due to their inherently rigid nature. This is mainly due to the fact that they can create cracks with flexible substrate when the device is bent, so in order to overcome these problems, organic electrode materials are beginning to receive attention from researchers.

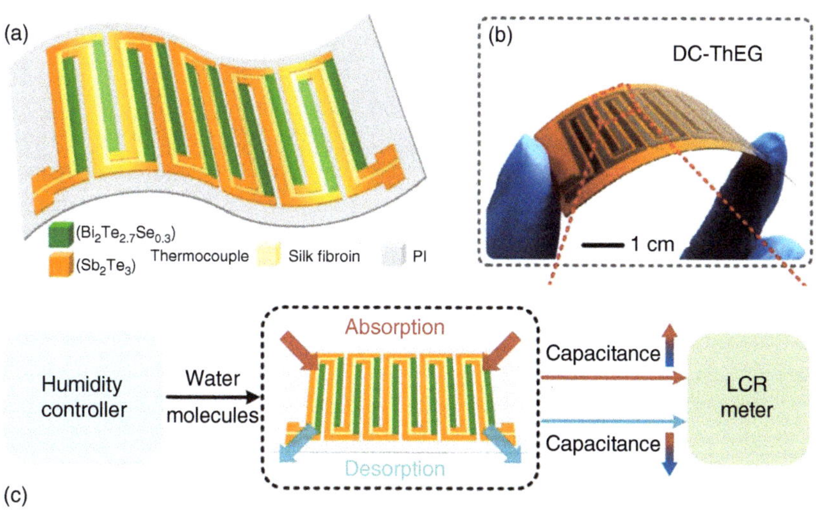

Figure 6.8 Schematic illustration and photo images of TEG. Source: Wen et al. [24]/Springer Nature/CC BY 4.0. (a) Schematic diagram of the TEG, (b) photograph of a bent TEG, (c) functional sensing mechanism.

Elmoughni et al. mixed and dispersed poly(3,4-ethylenedioxythiophene) polystyrene sulfonate (PEDOT:PSS) with a dimethyl sulfoxide (DMSO) solution and ground it in a micro-vibrating dispersion for 15 minutes to obtain a p-type ink. The n-type ink was obtained by mixing poly[Na(NiETT)] ink with polyvinylidene fluoride (PVDF)/DMSO solution. Subsequently, 32 holes were cut out of the prepared polyester fabric, and the p- and n-type inks were alternately deposited into the holes using a screen printing technique, the same procedure was repeated on the reverse side and finally the TE legs were joined together using a silicon-based silver ink. The device can produce 3 mV at a temperature difference of 3 °C [26].

Zhang et al. innovated the application of roll-to-roll (R2R) printing to the preparation of TEG devices, where p-type inks are made from PEDOT:PSS dispersed in 5% DMSO and n-type inks are made from nitrogen-doped graphene sonicated into ethylene glycol (EG) to form a 0.1 mg ml^{-1} solution, both of which can be stable for more than 12 hours. The R2R printing system consists of three continuous cylinders that transfer the ink onto the substrate by means of adjustable rotational pressure and speed. The PEDOT:PSS ink then joins the TE leg structures together. It is worth noting that the contact angle between the substrate and the ink is significantly reduced, after ultraviolet irradiation and oxygen plasma treatment. The device is cut into a rectangular film of 8 cm × 10 cm with an output power of ~0.24 mW m^{-2} at $\Delta T = 10$ °C [27].

Zhang et al. constructed thin film-like TEGs by improving R2R printing of metals. p-type and n-type inks were both made from single-walled carbon nanotubes (SWCNTs)/poly(3,4-ethylenedioxythiophene) (PEDOT):PSS and SWCNT/triphenylphosphine, and the preparation steps were divided into the following four main steps, electrodes printing, lamination process, TE ink PDMS is printed onto a polyimide (PI) substrate. After drying, the PDMS film is perforated and pressed onto the web at 20 mm intervals. Afterward, using inkjet printing, p- and n-type inks are filled in the holes to create electronic pathways, and after the inks have cured, the cavities are encapsulated with PI film. This method can effectively increase the efficiency of TEG device preparation, and the device shown in the figure is 152.4 cm wide and can reach several meters in length, making it an example of printing at scale. Devices consisting of 20 PN junctions can output a maximum voltage of ~20 mV and exhibit a maximum output power of ~6.5 μW at a temperature gradient of 25 K [28].

In order to develop an innovative unipolar TEG device, Wang et al. used a SWCNT/polyaniline (PANI) solution to configure a p-type ink and used mask masking to construct a strip pattern by dropping the compliant solution onto the substrate. Subsequent drying at 408 °C gave p-type legs measuring 22 mm × 10 mm × 0.15 mm. Silver paste and gold wire were used as connecting wires to join the legs together. The composite film could achieve a maximum in-plane conductivity and power factor of 1.44 × 103 S cm^{-1} and 217 μW m^{-1} K^{-2}. Furthermore, this TEG device can contribute 0.012 μW cm^{-2} at a temperature difference of 30 °C [29] (Figure 6.9).

Zhou proposed a strategy for the construction of TEGs based on the alternate conversion of p-type films to n-type films. First, the authors used a floating catalyst chemical vapor deposition (FCCVD) method to synthesize pristine SWNT films,

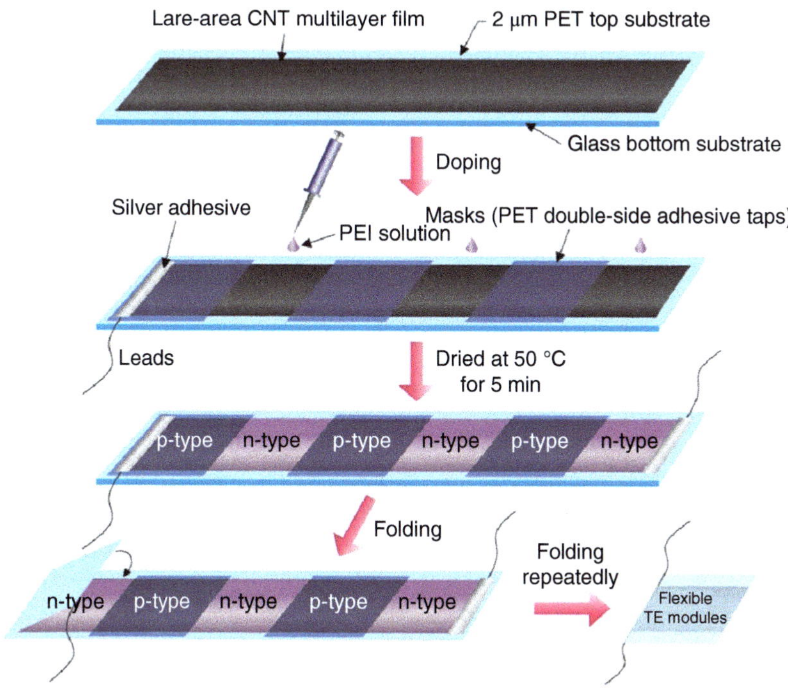

Figure 6.9 Schematics of the fabrication process for TEG device. Source: Zhou et al. [30]/Springer Nature/CC by 4.0.

which were subsequently transferred to PET substrates as p-type films. The SWNT films exhibited a power factor of up to 1840 $\mu W\,m^{-1}\,K^{-2}$ due to the doping of oxygen in air to create cavity-like carriers. Then the polyethylene terephthalate was used to mask parts of the films to form strip patterns. Drops of ethanol solvent with different concentrations of polyethyleneimine (PEI) were configured onto the films. At low PEI concentrations, only a small number of electrons are doped, causing the dielectric constant of the film to drop. As the PEI concentration rises to 1 wt%, the Seebeck coefficient of the doped SWNT tape becomes negative, indicating n-type properties. After removing the tape, the final TEG device is obtained by repeatedly folding the film. A maximum output power of 167 $\mu W\,cm^{-2}$ at $\Delta T = 27.5\,K$ can be produced when the device size is 16 mm × 10 mm × 0.15 mm. More importantly, the device can remain stable in air for long periods of time, with a change in conductivity and Seebeck coefficient of less than 5% over a period of three months. The strategy is simple and easy to implement, the devices prepared have excellent performance and the upcoming industrial production offers opportunities [30].

Stephan et al. propose the method of using dispenser printing to create p-type electrodes on glass. The PEDOT:PSS ink is dispensed sequentially onto the substrate via a nozzle, a strategy that avoids the use of mask shading and simplifies the production process. The TE legs are then joined together using a silver paste to form a flexible thermoelectric TEG. The results show that the device can deliver a maximum output power of 0.025 $\mu W\,cm^{-2}$ at a temperature difference of 90 °C [31].

Similarly, Liu et al. used PDMS printed on a PET base as a mask, followed by electrodeposition of poly[Kx (Ni-ethylenetethionate)], thereby enabling the modulation and preparation of the pattern. The method has been experimentally validated to allow the preparation of 108 n-leg at a time. The electrodes on the TE films were subsequently connected with gold to construct a complete TEG device, which had a maximum output power of 577.8 μW cm^{-2} at a temperature difference of 12 °C [32].

Zeng et al., first used 3D printing to construct a 3D lattice structure of PDMS, connected with a Cu/Ni-coated conductive fabric as a base. To enhance the hydrophilicity of the PDMS, a very thin copper film was coated on its surface. It was then placed in a graphene dispersion solution. A continuous rGO film is then formed on the copper-coated PDMS grid. This side rate effectively maintains the unique 3D-printed lattice structure supporting the rGO film and reduces unnecessary heat loss increasing the output power. Finally, the TEG is covered with another layer of conductive fabric on top of the TEG. The TEG has good instrumental strength and flexibility and did not show any breakage after 100 bends, and a single device (10 mm × 10 mm × 2 mm) produced an output voltage of approximately 57.33 mV g^{-1} at $\Delta T = 50$ °C. To add to this, the authors integrated the seven units developed with one side attached to the skin of the wrist and the other side exposed to air. An output voltage and output power of 1.47 mV and 0.88 μW were generated at an impedance-matched external load of 2.44 Ω at an air temperature of 15 °C (10 days) [33] (Figure 6.10).

Wu and Hu synthesized a fibrous p- and n-type thermoelectric electrode using a nonionic aqueous polyurethane (NWPU)/PEDOT: PSS/MWCNT composite with NWPU/nitrogen-doped MWCNT wrapped separately on a commercial polyester yarn. To fabricate the device, the fibrous electrodes were embroidered into a spacer fabric substrate and arranged alternately. Conductive coatings are added to the top and bottom of adjacent p- and n-type yarn loops to connect each thermoelectric leg, forming a final TEG device. Such a 3D device uses air as a filler in the middle and a temperature difference is created between the top and bottom layers. The device consists of two separate fabric layers with good compressibility, thermoregulation, and a wide range of structural variations. When demonstrated as a device with a human skin temperature of 33.2 °C and a room air temperature of 26.8 °C, a thermal voltage of 0.025 mV can be generated, while extending the thermal difference to 10.7 °C results in an output voltage of 0.1242 mV [34].

Lund has further developed this fiber-weaving process. The author uses an aqueous PEDOT:PSS dispersion mixed with ethylene glycol and immerses the silk sewing thread in it. The coated thread is then placed in a solution containing dimethylsulfoxide for one hour and 20 minutes and finally dried with a hot air gun at 100 °C in air to obtain the p-type fiber. For the n-type fibers, a silver-plated polyamide wire is used. Next to construct the device, the conductive thread will be threaded onto a sewing needle and sewn through nine layers of felt wool fabric, the thermocouple leg will be attached with a silver paste and the paste will be cured in an oven at 100 °C for 10 minutes. When the thermocouples are doped into the insulating wool fabric, the body constitutes a thermal reservoir and the external air a cold reservoir when worn in winter, allowing for both thermal insulation and differential heat

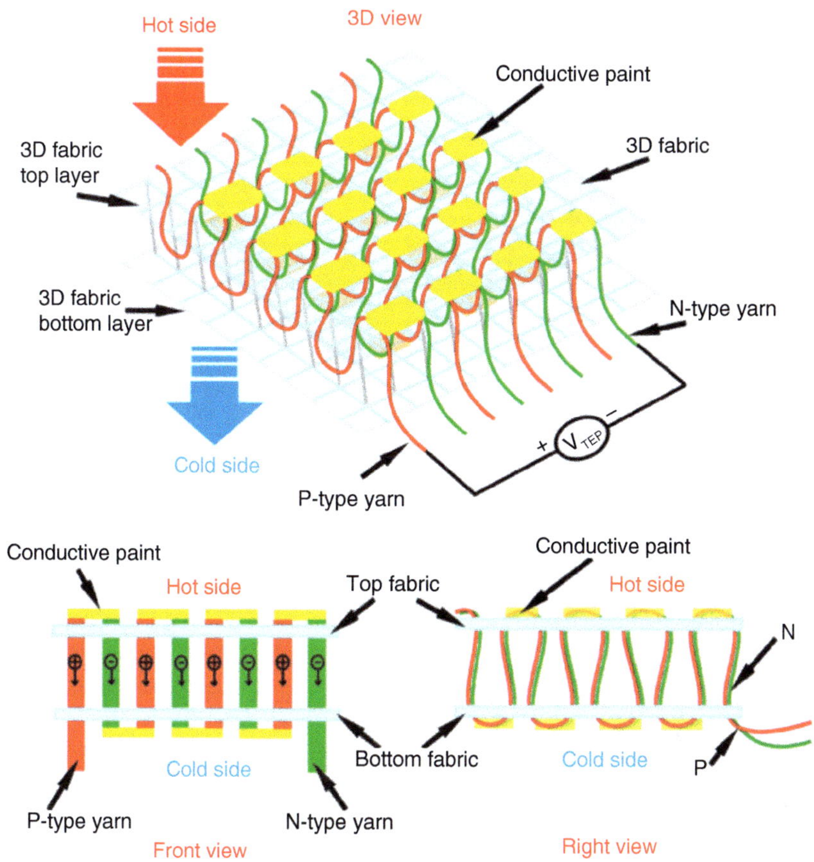

Figure 6.10 Structure design of the 3D fabric TEG. Source: Wu and Hu [34]/IOP Publishing.

generation. The device contains eight thermocouples achieving a maximum output power of 0.04 μW cm^{-2} at a ΔT of 65 °C [35].

Qu et al. instead proposed the use of a complete thread instead of separate p- and n-type threads. To achieve this strategy, p-type threads were constructed by immersing cotton threads in a poly(3-hexylthiophene) (P3HT)/toluene solution for 12 hours, followed by drying in an oven at 100 °C for 30 minutes and cooling at room temperature. To construct the n-thread, a portion of the P3HT wrapped around the thread is brushed with Ag as the n-thread material. This was then sewn onto a flexible fabric (thickness ~3 mm) with a distance of approximately 1 mm between them, thus obtaining a flexible TE device. The use of coated cotton threads allows the devices to be kept flexible and bendable, controllable, and can be prepared on a large scale. To test their thermoelectric performance, 13 thermocouples measuring ~4 mm long, ~1 mm wide, and ~1 mm thick, reached a maximum output power of 1.15 μW at a temperature gradient of ~50 K. Meanwhile, the authors face description came as a function of the thickness of the P3HT and Ag slurry coating, demonstrating that too thick a material coating reduces the actual temperature difference between the two sides of the wire, lead to reduce the power generated [36].

A similar idea was applied by Ito et al. and the p- and n-sections were electrically linked together. The carbon nanotubes/polyethylene glycol (CNT/PEG) composite line was obtained by injecting a mixture of polyethylene glycol and SWCNT into methanol as a coagulant and drying it in air after 24 hours. The CNT/PEG composite line was then obtained. The Seebeck coefficient is positive and exhibits few changes with different concentrations of PEG, but the conductivity decreases with increasing PEG, so the concentration of PEG was finally determined to be 0.01 wt%. As for the n-type electrodes, the fibers of CNT/PEG were n-doped using 1-butyl-3-methylimidazolium hexafluorophosphate ([BMIM]PF6) containing 10 wt% DMSO as a dopant. A complete p- and n-type wire is obtained by applying a conductive silver paste at their junctions. The device shows no significant change in resistance (less than 2%) over the 160 cycles of bending. Sewing this wire onto a thick felt fabric 3 mm thick with a base area of 400 mm^2, the TEG device achieved a maximum output power of 0.125 nW cm^{-2} at 22 °C ambient temperature and human body temperature (~37 °C) [37] (Figure 6.11).

Zheng et al. instead made segmented CNT fibers, forming complete p–n-type yarns. The original CNT was wound on thin (PET) sheets and rolled into a cylindrical structure. The p- and n-type legs were constructed by dipping them alternately into PEDOT aqueous: PSS and PEI/ethanol solutions, and finally 996 p–n pairs of TEG devices (8 cm × 9.3 cm 3.5 mm) with a transverse surface density of ~7.7 mm^2 were brought directly to the arm. The excellent bendability demonstrated a good fit to the human skin, which could be bent, twisted, and compressed at will without

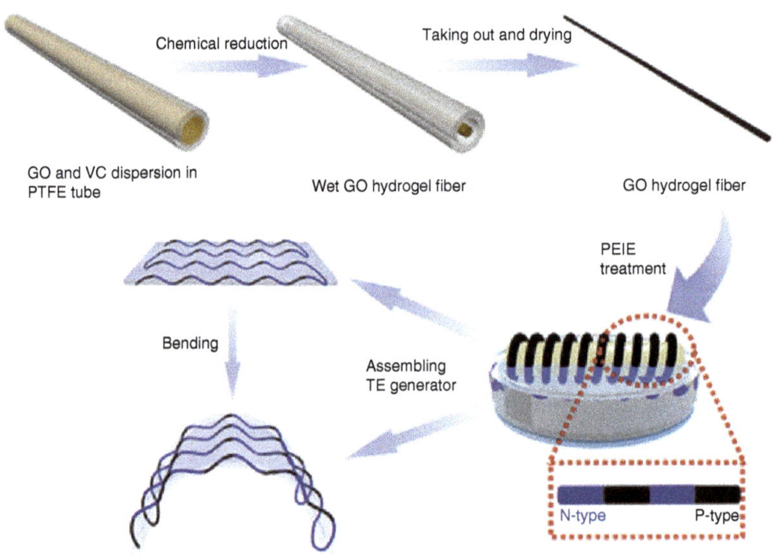

Figure 6.11 The schematic illustration of the fabrication process of an integral p–n connected all-graphene fiber. Source: Adapted from Ito et al. [37].

degrading the thermoelectricity. The maximum output voltage and power are 0.62 V and 10.12 μW at ΔT of ~7 K [38].

Lin et al. used PDMS to replace the commonly used 3D fabric substrate. p-Type graphene fibers were made by homogeneously mixing graphene oxide and ascorbic acid solution and then encapsulating them in a poly tetrafluoroethylene (PTFE) tube. After drying, the fibers were removed from the shutter and chemically treated with N_2H_4–H_2O to achieve higher TE properties. The p-type graphene fibers are then rolled onto plastic rods and partially immersed in the PEIE solution. Finally, the integrated p–n graphene fibers were attached to PDMS to obtain a flexible TE device. After treatment with N_2H_4–H_2O, it was possible to increase Young's modulus of the fibers by about 40% to 25.8 GPa, which facilitated the flexibility and mechanical stability of the devices. The TEG devices prepared using 20 pairs of p–n legs showed a change in resistance of approximately less than 5% during bending. Each p–n leg can deliver 0.75 mV and 124.1 nW at a temperature difference of 70 K. The maximum output power is approximately 1.3 pW at a difference of 10 ± 0.5 °C between body and air temperature [39] (Figure 6.12).

Sun et al. used four thin films of CNT to form CNT fibers to enhance the tensile strength of the devices. The CNT fibers consisted of ~280 μm in diameter and the interior consisted of multi-walled carbon nanotubes of ~11 nm in diameter. The CNT fibers were immersed in PEDOT:PSS solution for p-hybridization. The n-type CNT fibers were constructed using oleylamine doping combined with electro-spraying techniques with a polypropylene mask masking. After doping the fibers need to be

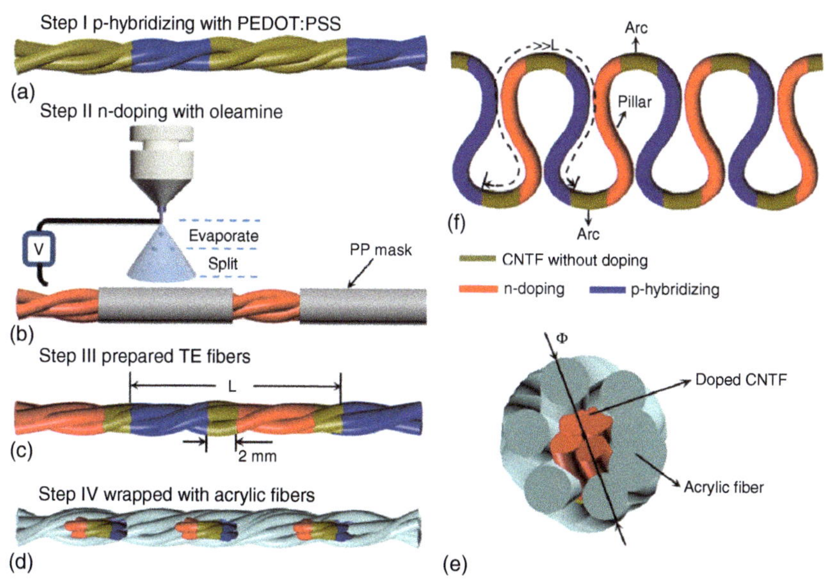

Figure 6.12 Schematic of the fabrication procedure of the TE fiber. Source: Sun et al. [40]/Springer Nature/CC by 4.0. (a) p-hybridized CNT. (b) n-doped CNT using an electrospray technique. (c) TEG fibers. (d) TEG fibres wrapped with acrylic fibers. (e) Cross sectional view for the structure of TEG. (f) TEG loops.

wrapped with acrylic fibers to avoid short circuits. These TE modules are woven in an alternating interlocking structure into a pattern of 3D TEGs without a substrate. These individual TEG modules exhibit excellent flexibility, with a change in thermal resistance of less than 0.5% in multi-angle bending tests. At the same time, the TE textile is not only flexible but also exhibits excellent tensile properties (~80% strain). The elastic force generated by the device's utilization of the locking mode prevents unwanted mechanical rupture of the device. According to the authors' speculations, when the TE garment covers ~40% of the surface area of an adult, ~200 μW of power, and ~20 V of open circuit voltage can be generated. Thus, promising applications for powering micro-watt wearable body sensors [40].

6.3.3 Organic–Inorganic Composite Material

The above article describes the application of inorganic and organic materials in flexible thermoelectric, where the inorganic material Seebeck coefficient is high, but there is a high thermal conductivity and brittleness, while the organic material cannot show enough power and its stability is poor but exhibit good film formation and high flexibility. Therefore, the combination of the two materials can effectively compensate for the shortcomings of both materials (Figure 6.13).

Bae took advantage of the excellent thermoelectric properties of Te nanorods to construct inorganic–organic composites using Te-PEDOT:PSS. However, the low electrical conductivity of this material cannot allow for direct application in thermoelectric devices. Therefore, the authors used H_2SO_4 chemical treatment to improve the properties associated with Te-PEDOT: PSS, with a significant increase in conductivity from an initial 11.01 S cm to 334.68 S cm^{-1} and a high-power factor of 284 μW m^{-1} K^{-2} due to a change in crystallinity caused by the structural rearrangement of PEDOT: PSS as a result of PSS removal. A thermoelectric generator using Te-PEDOT:PSS coated 32 legs arranged in two rows, where flexible PET (30 mm × 10 mm) was used as the substrate and silver paste was used to connect the individual legs. The device exhibits a maximum output power of 10.59 nW [41].

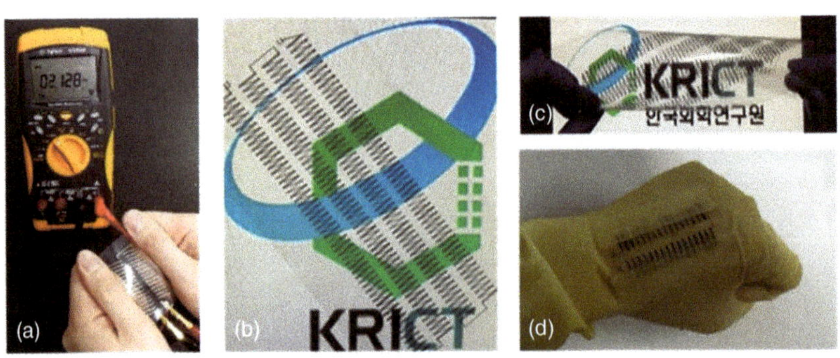

Figure 6.13 The photos of the flexible and twistable TEG. Source: Jin Bae et al. [41]/Springer Nature/CC BY 4.0. (a) Generation of electricity by grabbing both sides of the thermoelectric array at room temperature; (b,c) images of the flexible and twistable TEG and (d) demonstration of TEG embedded in a glove for the generation of electricity by human body heat.

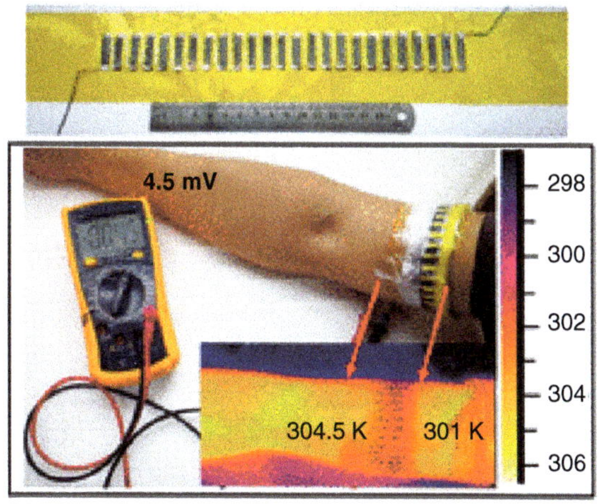

Figure 6.14 A photo of 4.5 mV voltage created from the TEG device. Source: Bae et al. [42]/Royal Society of Chemistry.

Subsequently, instead of continuing with the H_2SO_4 treatment of the material, Bae et al. used a partial carbon doping strategy. As a result, two common carbon materials, graphene nanoparticles (GNPs) and small bundled single-walled carbon nanotubes (SSWNTs), were added to the Te-PEDOT:PSS composite. The conductivity of SSWNT mixed with Te-PEDOT:PSS was significantly enhanced compared to that of GNP mixed with Te-PEDOT:PSS by experimental comparison. This is due to the longer conductive structure and electron transport path of SSWNT, which promotes carrier motion. The material was sprayed onto a glass plate as a TE component and tested thermoelectrically with 14 P–N legs, showing that a maximum output power of 126 nW was obtained with the 3 wt% SSWNTs-Te-PEDOT: PSS composite, far exceeding that of the 8 wt% GNPs-Te-PEDOT: PSS [42] (Figure 6.14).

Lu et al. grew polycrystalline Cu_xSe_y (PC-Cu_xSe_y) NWs *in situ* in PEDOT:PSS solutions, which were filtered and cold pressed to give thin films of PC-Cu_xSe_y. The *in-situ* growth approach allows the organic PEDOT:PSS and inorganic NWs to be firmly bonded together. The films are cut into 25 legs (25 mm × 5 mm) and attached to the polyimide substrate at ~5 mm intervals, electrically connected together with silver paste and finally protected with transparent tape. The device can exhibit a maximum output voltage of 4.5 mV at a temperature difference between the upper arm and the environment [43].

Subsequently, Lu further used *in situ* synthesis techniques to solve the problem of inorganic materials not being sufficiently dispersed in organic materials. Ag_2Se/CuAgSe NWs grown in situ in PEDOT:PSS can be obtained as PEDOT/Ag_2Se/CuAgSe composite films by vacuum-assisted filtration. The films, which can be used as n-type materials, were cut into 11 legs (25 mm × 5 mm) in a regular arrangement on PI plates and the nearest electrodes were joined together using a silver–silver adhesive. The authors also investigated the mechanism of material transformation during hot pressing. The gradual fusion of the PEDOT with the inorganic material led to a convergence of the lattice orientation and a resulting increase in electron mobility. The device showed a ~5% reduction in conductivity after 1000 bending cycles, demonstrating a good composite effect between the materials. When worn

Figure 6.15 The photos of the device and potential applications. Source: Wu et al. [45]/John Wiley & Sons.

on the arm, the TEG device can produce an output voltage of 6.7 mV with a temperature difference of (ΔT of ~5.3 K) [44] (Figure 6.15).

Wu et al. used RGO/Bi_2Te_3 as an n-type material and SWCNTs/Sb_2Te_3 as a p-type material to construct TEG devices. The authors found that the use of low-dimensional inorganic materials in combination with organic materials could greatly enhance the thermoelectric properties, mainly due to the planar orientation and network structure of the materials, as well as the synergistic application of vacuum filtration and annealing techniques. The TE film material is cut into uniform strips, alternately mounted onto polyimide tapes and electrically connected together with silver paste. To avoid short circuits and potential hazards from nanomaterials, the TEG was encapsulated using a flexible polyamide film. Experimental results show that TE devices can generate an open circuit voltage of 4.4 mV when used as wristbands with the skin as the heat source and air as the cold source. At the same time, the authors combined the TEG device with a flexible photovoltaic cell to effectively utilize the residual heat of the photovoltaic device and improve its efficiency and stability [45].

Xu et al. dissolved polyurethane (PU) in *N*,*N*-dimethylformamide to obtain a homogeneous mixture, which was applied to the pitch of the screw and dried naturally. Bi_2Te_3 as a p-type material and Bi_2Se_3 as an n-type material were then mixed with PVDF binder, respectively, to obtain a paste. The pastes were applied to the opposite side of the PU substrate and subsequently dried at 80 °C for 15 minutes to remove the organic solvent. A silver paste is used to join the gaps between the materials to ensure sufficient conductivity. After thorough drying, the material can be peeled off from the screw. Electron microscopy shows that the inorganic–organic materials are uniformly dispersed and tightly bound together, due to the penetration of PVDF into the PU substrate, which ensures a good interaction interface. The device remains stable at ~65% tensile deformation, with a slight drop in output voltage as the strain increases to 100%. After 1000 cycles of stretching, the TEG device maintains the same output voltage and output power. A stable open-circuit voltage of 8.9 mV can be generated at a temperature difference of ~16 °C between body temperature and room temperature [46] (Figure 6.16).

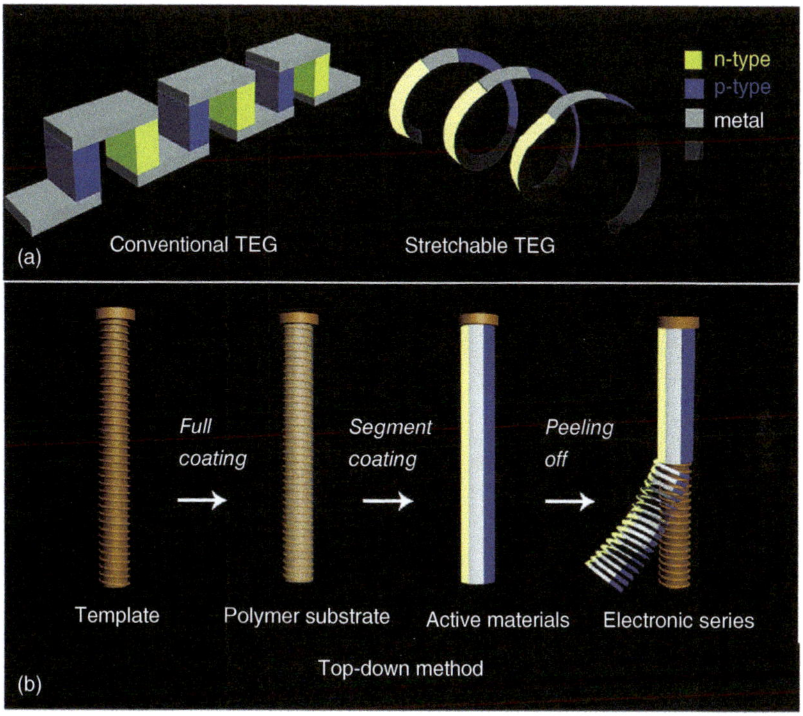

Figure 6.16 The schematic of the fabrication procedure of the TEG device. Source: Xu et al. [46]/Royal society of chemistry. (a) Comparison of the configurations of a conventional TEG and our stretchable helical TEG. (b) The highly stretchable, breathable and wearable based on the stretchable helical TEGs.

Highly aligned CNT sheets (200 mm × 15 mm) were synthesized using the FCCVD method by Choi et al. Also using $FeCl_3$ and PEI solutions doped at intervals to form p- and n-type legs with 2 mm spacing, other regions were connected to the legs using electrolysis. This continuous CNT structure facilitates the minimization of the circuit resistance of the module, resulting in high performance and mechanical stability. To verify the claim, the authors embedded nine contiguous p and n sections into the PDMS substrate to obtain a thermoelectric device that could be flexible. The device, repeatedly bent 1000 times with a 20 mm bending radius, also exhibits stable circuit resistance and achieves an output voltage of 3.4 mV at a ΔT of ~7 °C when it is worn on the wrist [47].

Some of these studies did not prepare composites but focused on inorganic or organic legs as p- or n-type legs, respectively, e.g. Na et al, used synthetic PEDOT and Bi_2Te_3 deposited to prepare as p- and n-type legs, respectively, and cut to 0.5 cm × 1.5 cm. Five pairs of p–n legs were prepared for adhesion to a flexible substrate, each piece measuring 5 mm × 15 mm and spaced 5 mm apart. Thermally evaporated gold (100 nm) was used as an electrode to connect the thermoelectric legs. The TEG so fabricated, after 200 repeated bending cycles, observed an increase in thermal resistance of 2.6% and could produce an output voltage of 0.93 mV when

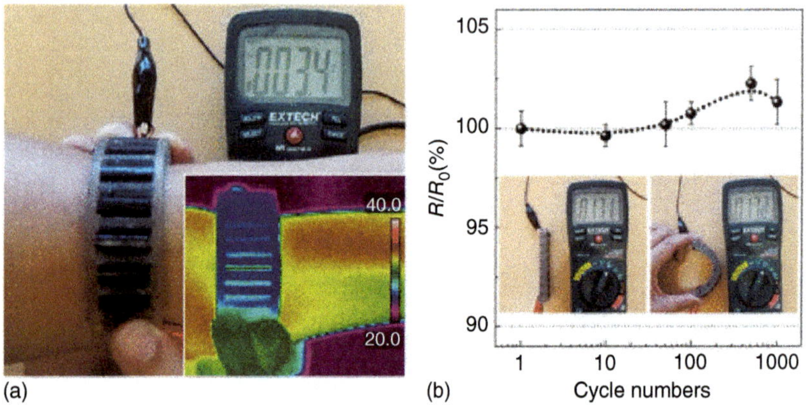

Figure 6.17 The photo of TEG device at the arm and binding test. Source: Choi et al. [47]/American Chemical Society. (a) Demonstration of biothermal energy harvesting from body heat. (b) Resistance changes as a function of bent cycles.

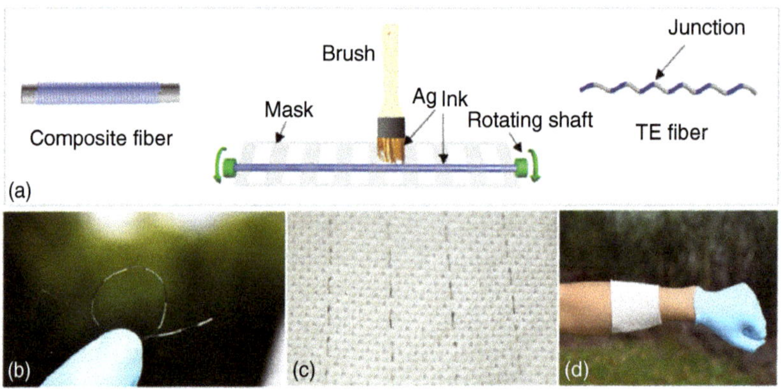

Figure 6.18 Schematic illustration of the fabrication of TE fibers. Source: Xu et al. [49]/American Chemical Society. (a) Schematic illustration of the fabrication of the fibers. (b–d) Photographs the application demo.

placed on the arm. A maximum power density of 3.45 nW cm^{-2} can be produced at a temperature difference of 12 °C [48] (Figures 6.17 and 6.18).

Xu et al. dispersed Te NWs in a solution of PEDOT:PSS at a concentration of 23 mg ml^{-1}. The dispersions were extruded in a coagulation and spinning technique using a wet process. The resulting fibers were placed in a sealed device containing 1 ml of ethylene glycol and left for two hours at 160 °C. The fibers were removed and dried and were ready for use in both p-type fiber electrodes. The silver paste is then brushed onto the PEDOT:PSS/Te fibers at equal intervals to act as n-type electrodes. To achieve long composite fibers, a small amount of PEG is added to the spinning solution. The performance parameters of the different lengths of NWs are compared, where the long fibers perform better than the short fibers due to the directional distribution of orientation in the long fibers and the formation of a continuous network

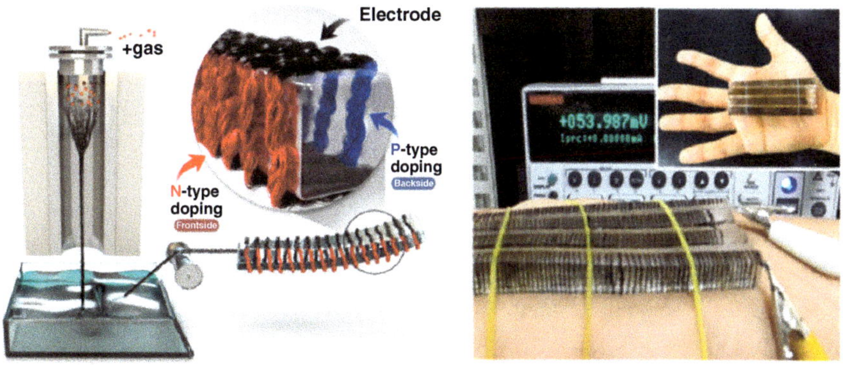

Figure 6.19 The doping process and photo of TEG device. Source: Choi et al. [50]/American Chemical Society.

of adjacent NWs. The treatment with glycol vapor increases the electrical conductivity of the fibers, although the Seebeck coefficient is only slightly reduced. Different Te contents have a large influence on the mechanical properties of the fibers, so that a Te content of 30% is the optimum ratio with a Seebeck coefficient, conductivity, and power factor of approximately 32 $\mu V\,K^{-1}$, 273 $S\,cm^{-1}$, and 28 $\mu W\,m^{-1}\,K^{-2}$, respectively. The final seven prepared fibers were stitched onto the preprepared textile as TEG devices and an output voltage of 8.3 mV was achieved at an ambient temperature difference of ~40 °C [49] (Figure 6.19).

Choi et al. synthesized carbon nanotube yarns (CNTYs) with excellent electrical conductivity of 3147 $S\,cm^{-1}$, due to the highly aligned oriented structure increasing the transport of electron carriers. The authors then wound the CNTY around PDMS blocks 4 × 10 × 80 mm and infiltrated them with $FeCl_3$ ethanol solution and PEI ethanol solution for 30 minutes on each side of the block, respectively, to achieve p- and n-type doping. The flexible TEG devices not only show outstanding mechanical strength but also have the advantage of modular fabrication in terms of shape. This is derived from the high tensile strength of CNTY, which is sufficient to support a weight of 5 kg and is comparable to some metals. The TEG consisting of 60 PN pairs has output voltage densities of 0.15 and 1.2 $V\,g^{-1}$ and maximum power densities of 10.85 and 697 $\mu W\,g^{-1}$ at temperature differences of 5 and 40 K. More notably, the device can increase the output power by increasing the number of PN pairs, while the average resistance of the increased PN pairs remains almost unchanged. Thus, a TEG device with 240 PN pairs can produce an output voltage of 54 mV with a temperature difference of approximately 6 K when covered with the arm [50].

6.4 Summary and Future Trend

Over the last few decades, micro-devices such as portable telephones, automotive pacemakers, and exoskeletons have placed greater demands on realizable energy sources. Thermoelectrochemistry, as a new field of research, has proven to be able to power these portable electronic devices by converting body heat into useful electrical energy, and therefore there is a huge potential for wearable applications, offering a wealth of opportunities for a wide variety of scientists. At the same time, at the very bottom of this field, there are many high-value questions worth exploring regarding the physics of carrier transport in organic and organic–inorganic thermoelectric systems.

This section focuses on a review of the important results achieved in this field in recent years, which can be briefly summarized as follows: (i) A variety of high-performance thermoelectric materials have been prepared through the processing of thermoelectric materials, the compounding of components and the optimization of synthesis conditions. (ii) A creative flexible thin film and fibrous energy conversion devices have been proposed with the structural design of thermoelectric devices. (iii) Utilizing the body and waste heat, heat lesions are successfully collected for the continuous power supply of the devices.

Despite these advantages that thin film and fiber-based materials have shown for flexible or wearable TE energy conversion devices, much more effort is needed in the development of many aspects for flexible or wearable TE devices. (i) Although many designs have been tried and made some progress for wearable applications, they are still far away from commercial applications. The utilization of multiple devices in series and parallel would increase the size of the devices. (ii) The capacity of scale production is poor. Although many thermoelectric devices have excellent performance, the design process is too complex to be practical applications. (iii) The wearable energy supplement would suffer stretching and bending when the human body is moving. Thus, the high demands on the mechanical strength and flexibility of the TEG film or fibers should be focused.

References

1 Bahk, J.-H., Fang, H., Yazawa, K., and Shakouri, A. (2015). *J. Mater. Chem. C* 3: 10362.
2 Dargusch, M., Liu, W.D., and Chen, Z.G. (2020). *Adv. Sci. (Weinh)* 7: 2001362.
3 Li, C., Jiang, F., Liu, C. et al. (2019). *Appl. Mater. Today* 15: 543.
4 Dai, B., Gao, C., and Xie, Y. (2022). *View* 3: 20220027.
5 Liu, E., Negm, A., and Howlader, M.M.R. (2021). *Mater. Today Energy* 20: 100625.
6 Nozariasbmarz, A., Collins, H., Dsouza, K. et al. (2020). *Appl. Energy* 258: 114069.
7 Siddique, A.R.M., Mahmud, S., and Heyst, B.V. (2017). *Renew. Sustain. Energy Rev.* 73: 730.
8 Suarez, F., Nozariasbmarz, A., Vashaee, D., and Öztürk, M.C. (2016). *Energy Environ. Sci.* 9: 2099.
9 Soleimani, Z., Zoras, S., Ceranic, B. et al. (2021). *Nano Energy* 89: 106325.
10 Tanwar, A., Lal, S., and Razeeb, K. (2021). *Energies* 14: 2339.
11 Zaia, E.W., Gordon, M.P., Yuan, P., and Urban, J.J. (2019). *Adv. Electron. Mater.* 5: 1800823.

12 Suarez, F., Parekh, D.P., Ladd, C. et al. (2017). *Appl. Energy* 202: 736.

13 Jung, S.-J., Shin, J., Lim, S.-S. et al. (2021). *Nano Energy* 81: 105604.

14 Hong, S., Gu, Y., Seo, J.K. et al. (2019). *Sci. Adv.* 5: eaaw0536.

15 Wang, Y., Shi, Y., Mei, D., and Chen, Z. (2018). *Appl. Energy* 215: 690.

16 Liu, H., Wang, Y., Mei, D. et al. (2017). Wearable Sensors and Robots. In: *Proceedings of International Conference on Wearable Sensors and Robots 2015*, pp. 55–66. Springer, Singapore.

17 Yuan, J. and Zhu, R. (2020). *Appl. Energy* 271: 115250.

18 Eom, Y., Wijethunge, D., Park, H. et al. (2017). *Appl. Energy* 206: 649.

19 Siddique, A.R.M., Rabari, R., Mahmud, S., and Van Heyst, B. (2016). *Energy* 115: 1081.

20 Varghese, T., Hollar, C., Richardson, J. et al. (2016). *Sci. Rep.* 6: 33135.

21 Guo, Y., Dun, C., Xu, J. et al. (2018). *ACS Appl. Mater. Interfaces* 10: 33316.

22 Zhao, J., Ren, H., Liang, Q. et al. (2019). *Nano Energy* 62: 94.

23 Ren, W., Sun, Y., Zhao, D. et al. (2021). *Sci. Adv.* 7: eabe0586.

24 Wen, D.-L., Deng, H.-T., Liu, X. et al. (2020). *Microsyst. Nanoeng.* 6: 68.

25 Karthikeyan, V., Surjadi, J.U., Wong, J.C.K. et al. (2020). *J. Power Sources* 455: 227983.

26 Elmoughni, H.M., Menon, A.K., Wolfe, R.M.W., and Yee, S.K. (2019). *Adv. Mater. Technol.* 4: 1800708.

27 Zhang, Z., Qiu, J., and Wang, S. (2016). *Manuf. Lett.* 8: 6.

28 Zhang, Z., Wang, B., Qiu, J., and Wang, S. (2019). *Manuf. Lett.* 21: 28.

29 Wang, L., Yao, Q., Xiao, J. et al. (2016). *Chem. An Asian J.* 11: 1804.

30 Zhou, W., Fan, Q., Zhang, Q. et al. (2017). *Nat. Commun.* 8: 14886.

31 Stepien, L., Roch, A., Schlaier, S. et al. (2016). *Energy Harvest. Syst.* 3: 101.

32 Liu, L., Sun, Y., Li, W. et al. (2017). *Mater. Chem. Frontiers* 1: 2111.

33 Zeng, W., Tao, X.-M., Lin, S. et al. (2018). *Nano Energy* 54: 163.

34 Wu, Q. and Hu, J. (2017). *Smart Mater. Struct.* 26: 045037.

35 Lund, A., Tian, Y., Darabi, S., and Müller, C. (2020). *J. Power Sources* 480: 228836.

36 Qu, S., Chen, Y., Shi, W. et al. (2018). *Thin Solid Films* 667: 59.

37 Ito, M., Koizumi, T., Kojima, H. et al. (2017). *J. Mater. Chem. A* 5: 12068.

38 Zheng, Y., Zhang, Q., Jin, W. et al. (2020). *J. Mater. Chem. A* 8: 2984.

39 Lin, Y., Liu, J., Wang, X. et al. (2019). *Compos. Commun.* 16: 79.

40 Sun, T., Zhou, B., Zheng, Q. et al. (2020). *Nat. Commun.* 11: 572.

41 Jin Bae, E., Hun Kang, Y., Jang, K.-S., and Yun Cho, S. (2016). *Sci. Rep.* 6: 18805.

42 Bae, E.J., Kang, Y.H., Lee, C., and Cho, S.Y. (2017). *J. Mater. Chem. A* 5: 17867.

43 Lu, Y., Ding, Y., Qiu, Y. et al. (2019). *ACS Appl. Mater. Interfaces* 11: 12819.

44 Lu, Y., Qiu, Y., Cai, K. et al. (2020). *Mater. Today Phys.* 14: 100223.

45 Wu, B., Guo, Y., Hou, C. et al. (2019). *Adv. Funct. Mater.* 29: 1900304.

46 Xu, X., Zuo, Y., Cai, S. et al. (2018). *J. Mater. Chem. C* 6: 4866.

47 Choi, J., Jung, Y., Dun, C. et al. (2019). *ACS Appl. Energy Mater.* 3: 1199.

48 Na, J., Kim, Y., Park, T. et al. (2016). *ACS Appl. Mater. Interfaces* 8: 32392.

49 Xu, H., Guo, Y., Wu, B. et al. (2020). *ACS Appl. Mater. Interfaces* 12: 33297.

50 Choi, J., Jung, Y., Yang, S.J. et al. (2017). *ACS Nano* 11: 7608.

7

Thermoelectric Ionogel for Low-Grade Heat Harvesting

7.1 Introduction

Thermal energy is one of the essential energy sources in our production and living process, including geothermal energy, solar heat, industrial waste heat, and body heat. Thermal energy is an excellent secondary energy source and also a renewable energy source, which can be effectively converted into electrical energy and utilized effectively. The quality of thermal energy is evaluated according to the change of temperature. High-grade heat is a kind of energy that can be utilized with high efficiency, and it is the main force of thermal power generation at present. In addition, low-grade waste heat (below 100 °C) is very valuable as a broader thermal energy. Although the use of low-quality heat is more difficult, it is still possible to convert it into electricity to be collected and used. In particular, low-grade waste heat has the advantages of rich content, strong continuity, and wide distribution compared to high-quality thermal energy. The conversion of low-grade waste heat and electrical energy is rapidly developed in the fields of Internet of Things (IoTs), wearable electronics, electronic skin, and medical treatment (Figure 7.1). For example, the difference between body temperature and ambient temperature is typical of low-grade waste heat. This heat-to-electricity conversion device is a typical low-cost portable generator. In addition, devices with both low-grade waste heat-electrical energy conversion and storage have been continuously studied.

Currently, thermoelectric materials have two heat conversion mechanisms that can more efficiently convert low-grade heat into electricity. These two mechanisms can achieve ionic Seebeck coefficients that are many orders of magnitude higher than conventional conversion mechanisms. One mechanism is based on a faradaic redox reaction between hot and cold electrodes under a temperature gradient, known as the thermoelectric effect. The other is the thermal diffusion effect, called the Soret effect, which is similar to the Seebeck effect in semiconductors, and the Soret effect is also based on carrier diffusion (Figure 7.2) [1]. The thermoelectric materials with two conversion mechanisms at the same time exhibit a conversion efficiency far exceeding that of materials with a single-conversion mechanism.

Thermoelectric devices based on the Seebeck effect and Soret effect are particularly suitable for low-level heat conversion and storage. In the process of heat transfer from the high-temperature side to the low-temperature side, the potential

Low-Grade Heat Harvesting: Materials, Devices, and Technologies, First Edition. Xiaogang Zhang.

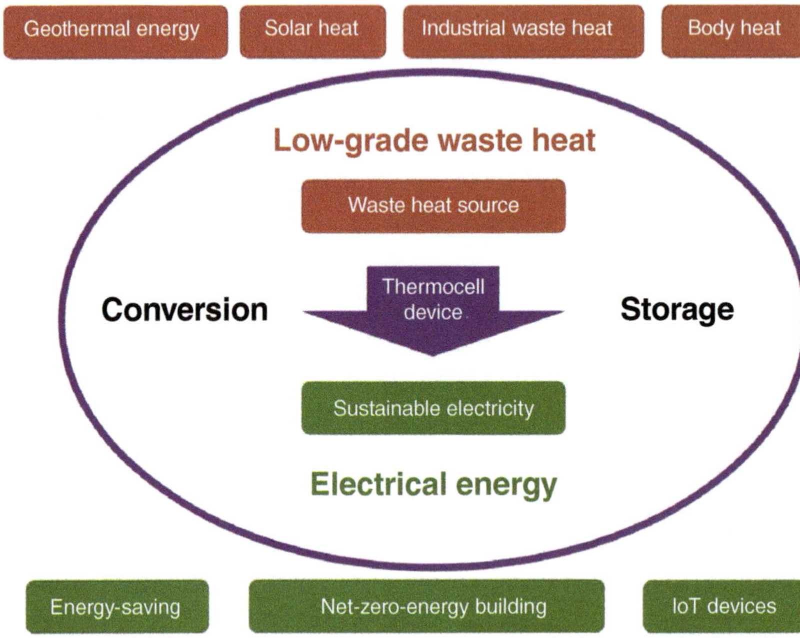

Figure 7.1 Heat resources and heat-recovering technologies can be used for sustainable power generation. Low-grade waste heat-electrical energy conversion, storage, and application.

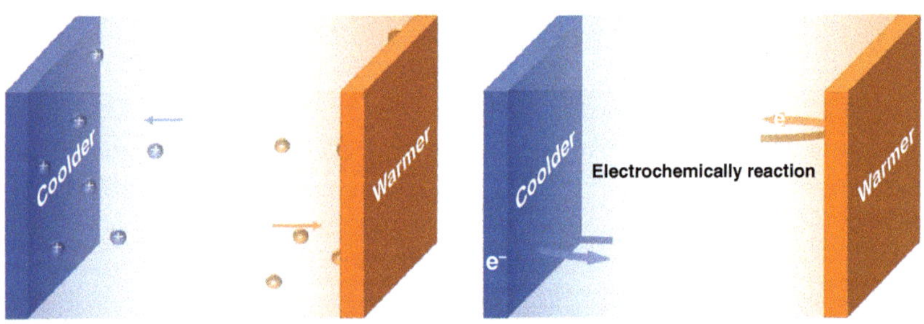

Figure 7.2 Schematic diagram of Soret effect and Seebeck effect.

difference generated on both sides of the device realizes the supply of electric energy to the outside world, which is due to the directional movement of internal carriers. Traditional thermoelectric materials are mainly inorganic semiconductors, semimetals, and conductive polymer materials, which rely on holes and electrons as carriers for thermoelectric conversion. Conventional inorganic semiconductors and semimetallic thermoelectric materials have the disadvantages of scarce elements, high fabrication costs, and complex synthesis. In addition, inorganic thermoelectric materials have little flexibility and stretchability, and their brittleness hinders their applications in the flexible field. Organic conductive polymer thermoelectric materials have the advantages of easy preparation, low cost, low thermal conductivity, and excellent mechanical performance. Compared with liquid electrolytes in thermoelectric devices, organic thermoelectric materials have the advantages of packaging and integration in the process of further promoting thermoelectric development.

This chapter systematically introduces the typical organic conductive polymer thermoelectric material-thermoelectric ionogels from three aspects. First, we briefly introduce thermoelectric ionogels and their performance parameters for thermoelectric applications. Then, we summarize the progress in design optimization and fabrication process of thermoelectric ionogels. At the same time, we pointed out the application and possibility of different thermoelectric ionogels in different fields. Finally, we discuss the challenges and some future approaches to improve the performance of low-grade heat-harvesting thermoelectric ionogels.

7.2 Thermoelectric Performance of Thermoelectric Ionogels

As a conductive polymer, thermoelectric ionogels need to meet two conditions at the same time to exhibit conductive properties. (i) The molecular orbitals of the molecular chains in ionogels can be strongly delocalized, which can generate a sufficient number of charge carriers. (ii) Conductive channels need to be formed within and between molecular chains of ionogels to provide carrier mobility. The conduction mechanism of ionogel polymers is different from that of thermoelectric materials including metals and semiconductors. The carriers of semiconductors are electrons or holes. The carriers of ion gels are the transport of ions by hopping within and between polymer molecular chains.

For thermoelectric materials, the three most critical parameters to judge their performance are electrical conductivity, Seebeck coefficient, and thermal conductivity. Currently, the design of ion thermoelectric systems is mainly based on the Soret effect and Seebeck effect. According to different mechanisms, ion thermoelectric systems can be divided into three models, including thermal diffusion batteries, pyrogen batteries, and a combination of both. In this section, we will point out the determinants of the performance parameters of ion gels while combining the transport characteristics of ion gels based on different thermoelectric transport mechanisms and energy storage mechanisms.

7.2.1 Basic Performance

Currently, the thermoelectric effect and energy conversion efficiency are the key factors affecting the performance in thermoelectric conversion and storage systems.

In terms of thermoelectric effects, there are mainly three basic thermoelectric effects that can explain the direct conversion behavior between electricity and temperature gradients, namely, Seebeck effect (Figure 7.3a), Peltier effect (Figure 7.3b), and Thomson effect (Figure 7.3c). The Seebeck effect, as the first discovered thermoelectric effect, was proposed by German physicist Thomas Johann Seebeck. It states that a nearby closed-loop joint of two dissimilar materials can sense a compass when

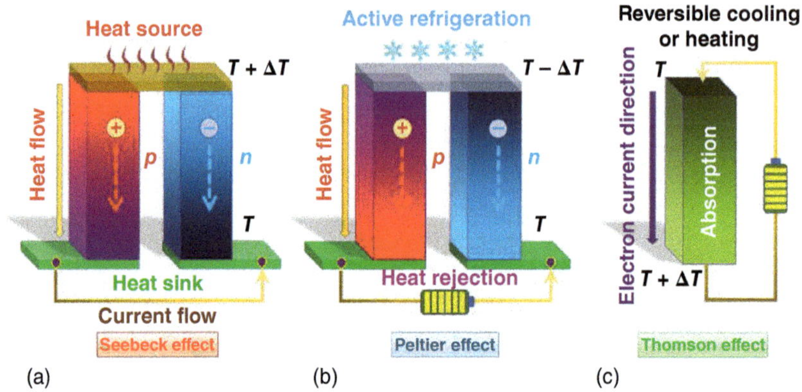

Figure 7.3 Schematic diagrams of thermoelectric effects. Source: Shi et al. [2]/American Chemical Society. (a) Seebeck effect, (b) Peltier effect, (c) Thomson effect.

a temperature gradient ΔT is loaded between the junctions. The relative Seebeck coefficient (S_{ab}) between two materials (a and b) can be expressed as follows.

$$S_{ab} = S_a - S_b = \lim_{\Delta T \to 0} \frac{V}{\Delta T} \tag{7.1}$$

where S_a and S_b are absolute Seebeck coefficients of materials a and b, and V is the thermoelectric force. The voltage formed by the Seebeck effect is derived from the asymmetric distribution of electron transport under the temperature gradient, which is the basic effect of thermoelectric power generation. The Peltier effect is when an electric current passes through a thermocouple, producing a slight cooling or heating effect depending on its direction. It was discovered by French physicist Jean Charles Athanase Peltier and can be described by the following equation.

$$\frac{dQ}{dt} = \pi_{ab} I = (\pi_a - \pi_b) I \tag{7.2}$$

where π_{ab} is the relative Peltier factor of materials a and b, π_a and π_b are the absolute Peltier factors of materials a and b, Q is the heat energy, t is the time of current passing, and I is the electric current. The Peltier effect is the fundamental effect of thermoelectric coolers. The mechanism is that the energy transferred by the electrons changes when an electric current passes between the materials. The Thomson effect is the reversible cooling or heating of a homogeneous conductor when current and temperature gradients coexist, proposed by William Thomson of Ireland. Thomson factor (τ) can be described as follows.

$$dQT = \frac{\tau I dt}{dx} \tag{7.3}$$

The Thomson effect is often neglected in thermoelectric device design and energy conversion analysis. The contribution to energy conversion is small compared to the Seebeck effect and the Peltier effect. However, the Thomson effect cannot be ignored when there is a large temperature gradient. According to the formulas of the three effects, the relationship between them can be expressed as follows.

$$\pi_{ab} = S_{ab} T \tag{7.4}$$

$$\frac{dS_{ab}}{dT} = \frac{\tau_a - \tau_b}{T} \tag{7.5}$$

Another key performance of thermoelectric conversion and storage systems is energy conversion efficiency. There are different expressions of energy conversion efficiency for thermoelectric generators and thermoelectric coolers. The power generation efficiency η of a thermoelectric generator is described by the following equation.

$$\eta = \frac{T_h - T_c}{T_h} \frac{\sqrt{1+Z\overline{T}} - 1}{\sqrt{1+Z\overline{T}} + \frac{T_c}{T_h}} = \varepsilon_c \frac{\sqrt{1+Z\overline{T}} - 1}{\sqrt{1+Z\overline{T}} + \frac{T_c}{T_h}} \tag{7.6}$$

where T_h and T_c are the hot-side and cold-side temperatures, $\overline{T}$ is the average temperature of T_h and T_c, and ε_c is the Carnot efficiency. The coefficient of performance of thermoelectric cooler is described by the following equation.

$$\text{COP} = \frac{T_h}{T_h - T_c} \frac{\sqrt{1+Z\overline{T}} - \frac{T_c}{T_h}}{\sqrt{1+Z\overline{T}} + 1} \tag{7.7}$$

In Eqs. (7.6) and (7.7), $Z\overline{T}$ is calculated by the following equation.

$$Z\overline{T} = \frac{\int_{T_c}^{T_h} ZT\mathrm{d}T}{T_h - T_c} \tag{7.8}$$

7.2.2 Thermodiffusion Cell

As mentioned above, the Soret effect refers to the phenomenon that different ions in the solution have different mobilities under temperature differences, resulting in the existence of ion concentration gradients in the electrolyte solution under temperature gradients. Similar to the Seebeck effect of electrons and holes in solid-state thermoelectrics, the temperature gradient (∇T) drives the migration of ions in the electrolyte to form a concentration gradient (∇c). S_{td} is the Soret coefficient, which can be denoted as follows.

$$\frac{\nabla c}{c} = -S_{td}\nabla T \tag{7.9}$$

Once the temperature gradient field is formed, the liquid phase in a thermoelectric device is divided into a dense ion region near the cold electrode and a loose ion region with high ion mobility near the hot electrode, resulting in an asymmetric ion concentration distribution between the two electrodes. In these thermoelectric materials, high-mobility ions tend to occupy the surface of the cold electrode, and then repel electrons from the cold electrode through Coulomb interactions, generating a thermovoltage. As the electrodes are electrically connected through the load, the thermal voltage causes electrons to flow in the circuit and produces an electric current.

The thermoelectric potential of a thermal diffusion cell depends on the difference in thermal mobility between anions and cations (Figure 7.4). However, the

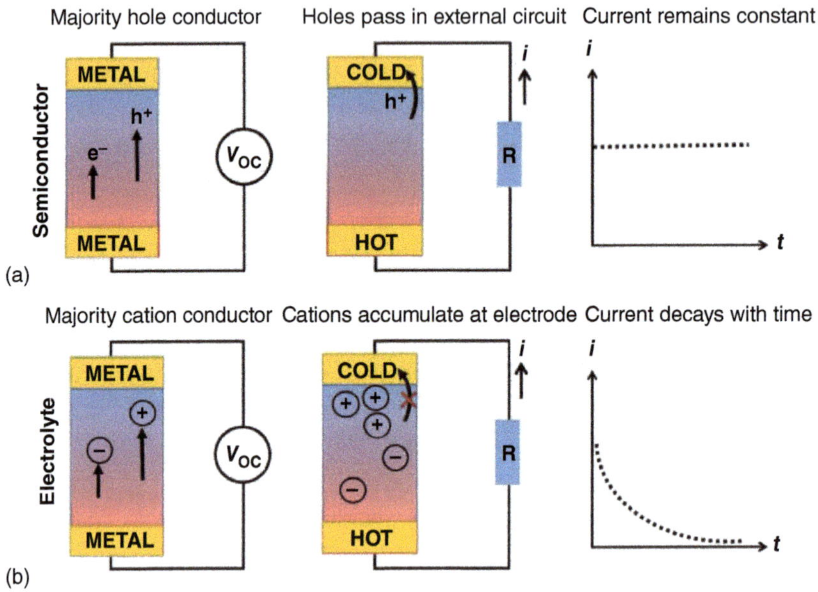

Figure 7.4 Thermoelectric effects with (a) electronic charge carriers in a semiconductor and (b) ionic charge carriers in an electrolyte. Source: Massetti et al. [3]/American Chemical Society/CC BY 4.0.

thermoelectric potential of liquid heat diffusion cells such as molten salts and ionic liquids is usually lower than 1 mV K^{-1} due to the small difference in thermal mobility of cations and anions. To increase the thermoelectric potential, one strategy is to select cations and anions with inherent diffusivity differences. Various ionic systems exhibit different thermoelectric potentials in their intrinsic states, depending on ion size, ionization capacity in solvents. For example, imidazole ionic liquids can achieve a larger thermopower due to the hydrophilicity and ion size difference. Another strategy is to break the local charge neutrality by confinement, which can generate unipolar thermal diffusion thermoelectric potentials.

Incorporating ionic electrolytes into rationally designed polymer frameworks yields ionogels that not only enhance thermoelectric potential but are also suitable for wearable applications. Selective interactions between ionic liquids and polymers with polar or charged functional groups amplify the thermal mobility difference between cations and anions, thereby enhancing the thermal diffusion thermoelectric potential. Currently, typical polymers are used in the gel framework of ion gels, including agarose, polyvinyl alcohol (PVA), polyvinylidene fluoride-*co*-hexafluoro-propylene (PVDF-HFP), polyacrylamide (PAM), gelatin, and cellulose.

Ionic gels are made from various ionic liquids and polymers. By comparison, they found that the thermoelectric potential is related to the ion–dipole interaction between the polymer and the ionic liquid, as shown in Figure 7.5c [4]. An ionogel made of [EMIM][DCA] and PVDF-HFP achieved a huge thermoelectric potential of 26.1 mV K^{-1}. Furthermore, by adjusting the composition percentages of PVDF-HFP

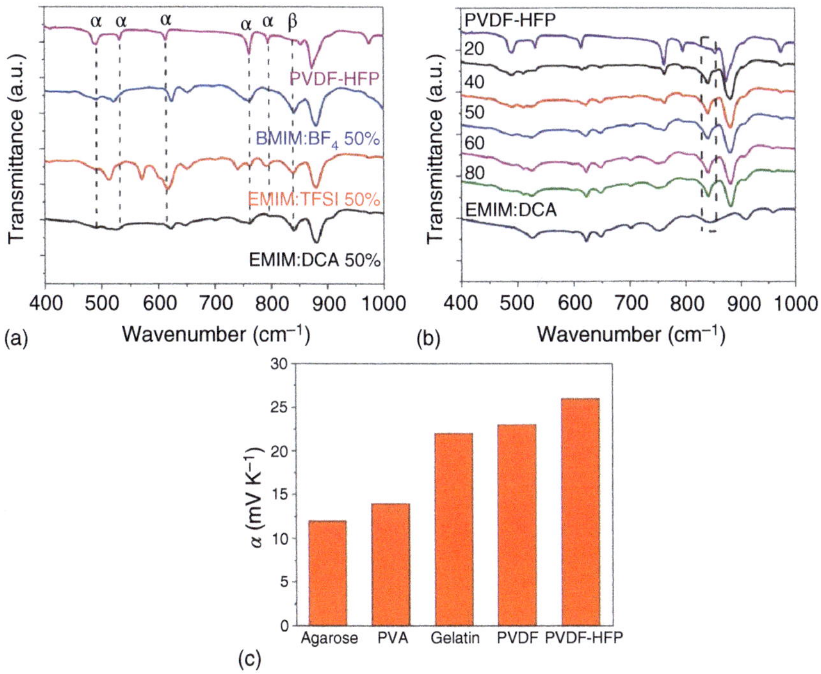

Figure 7.5 FTIR transmittance spectra of (a) PVDF-HFP and ionogels with different ILs and (b) PVDF-HFP, EMIM:DCA, and PVDF-HFP/EMIM:DCA ionogels with different IL loadings. (c) Seebeck coefficients of ionogels with the EMIM:DCA loading of 80 wt% with various polymers. Source: Cheng et al. [4]/John Wiley & sons.

and polyethylene glycol, the thermoelectric potential can be tuned in the range of −4 to 14 mV K^{-1} [5]. Polyethylene oxide (PEO) functionalized native cellulose to introduce oxidized and aligned cellulose molecular chains was used as the ionogel host [6]. Based on the strong hydrogen-bonding interaction between PEO and hydroxide ions, the cellulose ion conductor can trap hydroxide anions of NaOH solution. Furthermore, the subnanoscale confinement of ions enhances the selective diffusion of NaOH solutions under thermal gradients, resulting in a thermoelectric potential of 24 mV K^{-1}. In contrast, a thermoelectric potential of 11 mV K^{-1} was obtained using NaOH in PEO solution [7].

7.2.3 Thermogalvanic Effect

The thermogalvanic effect is the temperature dependence of the redox electricity on the existing standard electrode potential. The thermoelectric potential of a pyrogen cell follows the following equation.

$$S_{tg} = \frac{\partial E_\infty}{\partial T} = \frac{1}{nF}[(S_B + \widehat{S_B}) - (S_A + \widehat{S_A}) - n\overline{\overline{S_e}}] \tag{7.10}$$

where ∂E_∞ and ∂T are the potential difference and the temperature difference, respectively. n is the number of electrons involved in the redox reaction, and F is

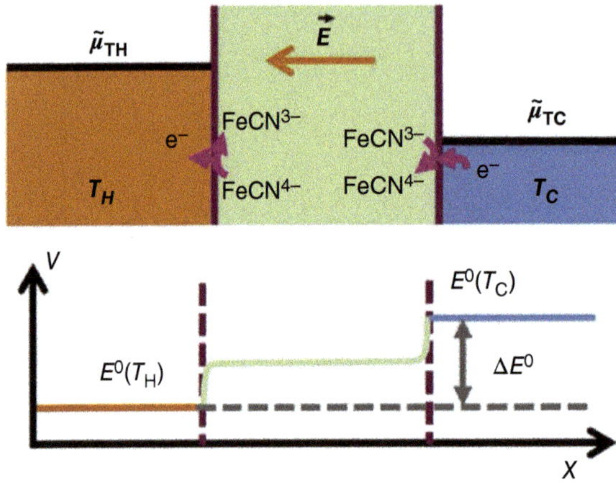

Figure 7.6 Energy and mechanism diagram describing the temperature-controlled redox reaction. Source: Liu et al. [8]/AIP Publishing.

Faraday's constant. S_A and S_B are the partial molar entropies of species A and B, while $\widehat{S_A}$ and $\widehat{S_B}$ are their Eastman entropies. $\overline{\overline{S_e}}$ is the transported entropy of the electrons in the external circuit.

$Fe(CN)_6^{4-}/Fe(CN)_6^{3-}$ is a typical redox couple for thermogalvanic batteries. In the initial state without thermal gradient, the redox process of the electrode is in equilibrium. There is no potential difference between the electrodes (Figure 7.6). When a temperature gradient is applied across a pyrogen cell, the temperature dependence of the redox reaction creates a potential difference across the electrodes. The potential difference created by the heat can drive electrons through an external circuit, converting the heat into electricity. Meanwhile, the concentration gradient of redox pairs between electrodes will drive the diffusion of these oxidized/reduced ions throughout the device to maintain equilibrium. As long as a temperature gradient is provided, a pyrogen battery can convert thermal energy into electrical energy, and its thermoelectric potential is on the order of $mV\,K^{-1}$. The thermoelectric potential of a pyrogen cell is related to the entropy difference of the redox couple. In liquid electrolytes, redox entropy can be estimated from the difference in solvation entropy between oxidized and reduced species. Different redox pairs have different entropy differences between oxidized and reduced species, resulting in different intrinsic thermoelectric potentials. The thermal conversion properties of some redox couples are shown in Table 7.1.

Liquid electrolytes usually have fast ion transport but are difficult to package, and there are certain safety issues. This is a challenge for wearable devices powered by body heat. Ionogel solid electrolytes are typical safe electrolytes that can be used in wearable applications. The gel-state electrolyte based on a cobalt redox couple using cellulose as the polymer framework was prepared, which exhibited a thermopower of $+1.15\,mV\,K^{-1}$ [12]. Using $Fe(CN)_6^{4-}/Fe(CN)_6^{3-}$ as a redox couple with poly(sodium acrylate) and PVA matrices to prepare ion gel electrolytes exhibited thermoelectric potentials of -1.09 and $-1.21\,mV\,K^{-1}$, respectively [14].

Table 7.1 The thermal conversion properties of some redox couples.

Redox couple	Thermopower	Measurement condition	Reference
$Fe/Fe(CN)_6{}^{4-}$	$-1.4\,mV\,K^{-1}$	Aqueous solution	[9]
$I^-/I_3{}^-$	$+0.97\,mV\,K^{-1}$	0.01 M aqueous solution	[10]
Iron (II/III)	$+1.02\,mV\,K^{-1}$	Common *n*-type redox couple	[11]
$[Co(bpy)_3]Cl_{2/3}$	$+1.21\,mV\,K^{-1}$	0.05 M $[Co(bpy)_3]Cl_{2/3}$ in water	[12]
$McMT^-/BMT$	$-0.6\,mV\,K^{-1}$	Organic redox couples	[13]
Cu^{2+}/Cu	0.84 mV	0.7 M aqueous solution	[10]

Using polyvinylpyrrolidone (PVP) as the polymer host with an $I^-/I_3{}^-$ redox couple to prepare a gel-state electrolyte, a thermoelectric potential of $+1.5\,mV\,K^{-1}$ was achieved. In contrast, the thermoelectric potential of the aqueous $I^-/I_3{}^-$ redox couple is $+0.86\,mV\,K^{-1}$. The improved thermoelectric potential of the $PVP/I^-/I_3{}^-$ gel-state electrolyte is induced by the introduction of polymer–ion interactions [10]. In addition, the incorporation of various ionic liquids is also a way to improve the thermoelectric potential and mechanical properties of gel-state pyrogen cells.

7.2.4 Synergistic Thermodiffusion and Thermogalvanic Effect

The limited efficiency of the thermoelectric conversion effect alone limits its practical application scenarios. An ionic gelatin was reported in which thermal diffusion effect and thermocurrent effect can work synergistically to enhance the overall thermoelectric performance [15]. Ionic gelatin obtained a huge thermoelectric potential of $17\,mV\,K^{-1}$. The thermal voltage of the synergistic thermal battery comes from the thermoelectric effect and thermal diffusion effect. The former is the thermodynamic shift of the chemical potential of redox species ($Fe(CN)_6{}^{4-}/Fe(CN)_6{}^{3-}$) due to temperature difference, and the latter refers to the selective thermal diffusion of ionic species (KCl) by the temperature gradient drive. The thermal power of the synergistic thermoelectric effect can be described by the following equation.

$$S_t = S_{tg} + S_{td} \tag{7.11}$$

7.3 Preparation of Thermoelectric Ionogel

Low-grade waste heat (below 100 °C) is widely generated throughout industrial or geothermal processes and is a particularly important source of energy that can potentially be harvested for power generation. Specifically, the fast developments in the area of IoT, wearable electronics, and e-skins have stimulated the emerging demand for self-powered supply in energy harvesting from the environment or human beings (Figure 7.7). Among recognized energy-harvesting technologies, such as solar cells and triboelectric and electric generators, low-grade heat is a promising power source

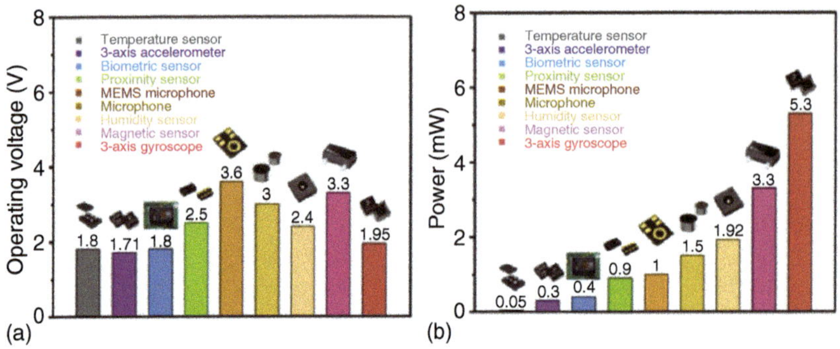

Figure 7.7 (a) Voltage and (b) power demands of some common IoT sensors. Source: Liu et al. [8]/AIP Publishing.

due to the wide availability in the environment, industrial processes, and the human body. For example, the human body has a surface temperature of about 32 °C, which is possible to form tens of degrees of temperature difference with the environment. In the human body, heat is continually being emitted (up to 20 mW cm^{-2}), which provides a constant power supply. In the past decades, research on harvesting such low-grade heat has increased significantly due to their ability to continuously convert heat into electricity without generating emissions or consuming materials.

Until recently, most of the research on thermal energy conversion was in the field of solid-state thermoelectric materials (e-TEs) using electrons or holes as energy carriers. However, thermoelectric devices based on semiconductor materials usually generate a relatively low thermopower (or Seebeck coefficient, Se) in the order of uV K^{-1}. Generally, IoT sensors require voltage of several volts and power from microwatts to megawatts. As a result, it is necessary but challenging to integrate thousands, or even tens of thousands, of tiny thermoelectric elements to meet the voltage requirements, limiting the practical development of e-TEs for wearable applications. Alternatively, ionic thermoelectrics (i-TEs) systems, which use ions as the energy carrier, have also been developed for direct low-grade heat harvesting, which typically exhibit much higher thermopower (on the order of mV K^{-1}) than e-TEs. Among them, thermogalvanic cells are based on the thermogalvanic effect, which can continuously generate electricity through redox-active electrolytes. Based on the Soret effect, ionic thermodiffusion cells operate in a capacitive mode, resulting in thermopower up to 26 mV K^{-1} under a temperature gradient without redox reaction. To prevent leakage, quasi-solid-state i-TEs have been obtained by adding gelation polymers into the liquid electrolytes. The resulting thermoelectric ionogel has great stability, durability, and deformability, especially suitable for wearable systems.

Thermodiffusion is the diffusion process of ions driven by a temperature gradient. Different ions in the solution have different mobility under the temperature difference, which leads to an ion concentration gradient in the electrolyte solution under the temperature gradient. This phenomenon is called the Soret effect. Similar to the Seebeck effect of electrons and holes in solid-state TEs, the temperature gradient

(∇T) drives the migration of ions in the ionic electrolyte, resulting in a concentration gradient (∇c) across the device. The Soret effect can be denoted as:

$$\frac{\nabla n}{n} = -S_{\mathrm{T}}\nabla T \tag{7.12}$$

where S_{T} is the Soret coefficient. In an ionic electrolyte, the different thermal mobility of cations and anions leads to a local net charge distribution, resulting in an electric field. Similar to e-TEs, the thermopower of i-TE is denoted as:

$$S_{\mathrm{td}} = \frac{\nabla V}{\nabla T} \tag{7.13}$$

where ∇V is the generated voltage of an i-TE system under temperature difference (∇T). According to Onsager transport theory, the S_{td} of the electrolyte is similar to the ambipolar diffusion

$$S_{\mathrm{td}} = \frac{n_+ z_+ D_+ \hat{S}_+ - n_- z_- D_- \hat{S}_-}{e\left(n_+ {z_+}^2 D_+ + n_- {z_-}^2 D_-\right)} \tag{7.14}$$

where the subscript $\pm$ represents cations (+) and anions (−), z is the valence of the ion, e is the element charge, and D is the particle diffusion coefficient. S is the amount of entropy transported with the ions flux, known as Eastman entropy of transfer, which is related to the interaction between the solute and the surrounding medium.

For a specific ion species, the Eastman entropy of transfer is associated with the Soret coefficient, as shown in the following equation:

$$S_{\mathrm{T}} = \frac{\hat{S}}{k_{\mathrm{B}} T} \tag{7.15}$$

where k_{B} is the Boltzmann constant and T is the absolute temperature. Once a temperature gradient field is created, the entire liquid phase of the i-TE material is divided into a dense ion area near the cold electrode and a loose ion area with high ion mobility near the hot electrode, resulting in an asymmetric ion concentration distribution between two electrodes. In these i-TE materials, the high-mobility ions tend to occupy the cold electrode surface, and then repelling the electrons of the cold electrode through Coulomb interaction, generating thermally voltage. As the electrodes are electrically connected by a load, the thermal voltage induces electron flow in the circuits and generates electricity.

Liquid thermoelectric electrolytes are not suitable in flexible and wearable devices. Recently, great progress has been made in leak-free quasi-solid gel-state electrolytes, with an aim for wearable thermocells. The term "quasi-solid" refers to the fact that there is generally a combination of solid polymer and liquid electrolyte, with solid-like mechanical properties and liquid-like ionic diffusion. In addition to being flexible and preventing leakage, gel-state electrolytes can also maintain a large thermal gradient throughout the cell by eliminating convection. In gel-state electrolytes, the selection of suitable polymer is important. In addition to providing a mechanical matrix sufficient to prevent solvent leakage, the polymer needs to maintain good ion diffusion. Sometimes it is even necessary to provide specific ion–polymer interactions to achieve ion selectivity. Different gelling agents have been used to make gel-state electrolytes, including cellulose, PVA, agar, sodium polyacrylate beads, gelatin, and so on. The chemical structures and functional groups of some widely studied gelling polymers are summarized in Figure 7.8 and Table 7.2. The balance between mechanical properties and thermoelectric performance such as thermopower and power outputs can be optimized by choosing suitable gelling.

Agents and adjusting the polymer/redox couple weight percentages. For example, when preparing the PVA-$FeCl_{2/3}$ gel-state electrolyte, a wide range of $FeCl_{2/3}$ concentrations (0.1–2 M) can be used without affecting the gelation process and maintaining the thermoelectric performance of the Fe^{2+}/Fe^{3+} redox couple. The mass diffusion coefficient of ions in ionogel is lower than that of ions in a liquid solution, causing a lower ionic conductivity. Reducing the thickness of the ionogel between the two electrodes can reduce the overall resistance of the thermocell, thereby increasing the output power.

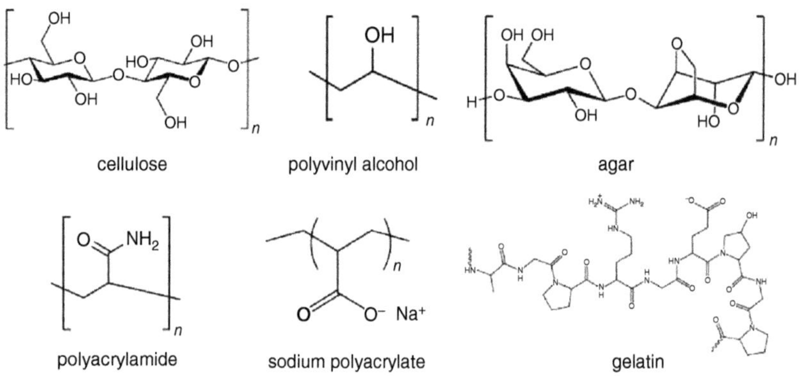

Figure 7.8 Gelling agent structures studied in thermoelectric ionogels.

Table 7.2 Summary of common gelling agents and gelation process.

Polymer	Functional group	Gelation process
Cellulose	–OH, hydrogen bond	Thermal dissolution and cooling gelation
Polyvinyl alcohol	–OH, hydrogen bond	Thermal dissolution and cooling gelation
Agar	–OH, hydrogen bond	Thermal dissolution and cooling gelation
Polyacrylamide	$-NH_2$, hydrogen bond	Polymerization and electrolyte soaking
Poly(sodium acrylate)	–COOH, $-COO^-$ negatively charged groups	Thermal dissolution and cooling gelation
Gelatin	$-COO^-$ negatively charged groups, $=NH_2^+$ positively charged groups	Thermal dissolution and cooling gelation

As for water-soluble polymers, such as agar, gelatin, poly(sodium acrylate), PVA, and celluloses, gelation is induced by cooling down a hot mixture containing electrolyte and gelling agent materials. In detail, homogenous solutions containing water or organic solvents with the specified weight percentage of gelling agent powder are prepared by stirring at high temperatures. The homogenous solutions are then poured into designed molds and allowed to cool to room temperature to obtain physically cross-linked ionogel with various shapes. It is also possible to immerse superabsorbent cross-linked polymers, such as PAM and cellulose, in a liquid electrolyte to produce ionogel. For example, to prepare PAM ionogel, cross-linked PAM can be synthesized by radical polymerization first. In detail, a homogeneous aqueous mixture containing AM monomer, cross-linking agent, and initiator can be cured in a designed mold. Then, the obtained hydrogel is soaked in liquid electrolyte for some time. After the complete solution exchange, an ionogel for thermocell applications can be obtained. For some water-insoluble gelling agent materials, ionogel can be prepared using organic solvents as an intermediate medium. First, the water-insoluble gelling agents can be dissolved in a suitable organic solvent to obtain physically cross-linked gels, and then the gel is immersed in the electrolyte after evaporating the organic solvent. For instance, HydroMed D series hydrogels were prepared by dissolving polymer granules in a mixture of ethanol (90 vol%) and water (10 vol%) at 65 °C, and then cooling down in a designed mold. After evaporating the ethanol/water solvent in the fume hood, the HydroMed D gel was placed in the electrolyte-containing solution until a fully swollen ionogel was formed.

7.4 Application of Thermoelectric Ionogel

As mentioned above, liquid thermoelectric electrolytes are not suitable for flexible and wearable devices and have safety issues. Leak-free quasi-solid gel-state electrolytes have made great progress and have been successfully used in various energy storage devices [16–18]. Therefore, flexible thermoelectric conversion devices also have great application prospects. "Ionogel" refers to the often existence of a combination of solid-state polymers and liquid electrolytes, with solid-like mechanical properties and liquid-like ion diffusion. In addition to being flexible and preventing leakage, gel-state electrolytes can maintain large thermal gradients throughout the thermal conversion device.

In ionogel electrolytes, suitable polymers are important. In addition to providing a mechanical backbone, polymers also need to maintain good ion diffusion. Sometimes it is even necessary to provide specific ion–polymer interactions to achieve ion selectivity. Different gelling agents have been used to prepare gel-state electrolytes.

Optimization between mechanical properties and thermoelectric properties (such as thermoelectricity and power output) can be achieved by selecting appropriate additives and adjusting the ratio of polymer and transport ion salt. In a wide range of concentrations (0.1–2 M), the PVA-$FeCl_2$/$FeCl_3$ ionogel electrolyte can maintain the thermoelectric properties of Fe^{2+}/Fe^{3+} without affecting the gelation process [19].

The mass diffusion coefficient of ions in ion gels is lower than that of ions in liquid solutions, which is the key factor leading to the lower ionic conductivity. By reducing the thickness of the ion gel between the two electrodes, the overall resistance of the thermoelectric conversion device can be reduced to increase the output power.

The thermal diffusion effect and thermal conversion effect in thermoelectricity have similar ion dynamics processes to the electric double layer ion adsorption energy storage and ion redox energy storage in ion capacitors. This also laid a solid foundation for the development of thermoelectric capacitors.

The thermoelectric conversion capacitor can be divided into four stages in the energy conversion process, according to the voltage and temperature difference distribution of the four stages. Yang and coworkers analyzed in detail the ionic gel thermoelectric capacitor constructed by 11 M Li_2SO_4 and 0.2 M $FeCN^{4-}/FeCN^{3-}$ [1]. Initially, the capacitor is discharged by shorting the positive and negative terminals with a copper wire. In this state, Li^+ and $SO_4{}^{2-}$ are uniformly distributed without diffusion in the gel. Meanwhile, $FeCN^{4-}/FeCN^{3-}$ maintains balance without redox process. It caused the thermoelectric device to generate no voltage. In stage I, after a temperature difference of 15 K is applied on both sides of the thermoelectric capacitor, the electrolyte ions undergo thermal diffusion and accumulate on both sides of the electrodes, finally forming an ion concentration difference inside the electrolyte. In addition, redox couples also undergo redox reactions on both sides of the electrodes. The thermal voltage generated at this time is the synergistic effect of thermal diffusion and thermocouple, and the maximum voltage value at this stage reaches 166 mV. In stage II, when a load resistor is connected in the capacitor, the accumulated and induced electrons flow rapidly from the hot side to the cold side. Likewise, this discharge current is also the result of the coupling of redox couples and ion diffusion. After the resistance is turned on, the electrons and holes in the circuit balance the difference in the concentration of anions and cations at the hot end and the cold end, and the voltage gradually decreases to zero. In stage III, the load resistance is disconnected, and the driving force of the temperature difference disappears, causing reverse diffusion of anions and cations on both sides. Due to the disconnection of the external circuit, the ions stay on the electrodes, thereby generating reverse voltage. In stage IV, after continuing to connect the load resistance, the previously accumulated polarized electrons and holes on the electrode are balanced by the external circuit. This stage produces a discharge voltage opposite to that of the second stage, and various ions in the gel will return to the discharge voltage. The initial state at the end. After these four processes are completed, various ions in the ion gel are continuously transferred to the opposite direction through diffusion, convection, and migration. It enables continuous power output between the two electrodes. Crispin and coworkers demonstrated for the first time that the Soret effect in polymer electrolytes leads to a significant i-TE effect for recharging supercapacitors (Figure 7.9) [7]. After removing the temperature difference, this thermal capacitor remains charged and can therefore be discharged by sending current to an external circuit. The demonstrated i-TE supercapacitor device requires fluctuations in the applied temperature gradient to function, making the concept suitable

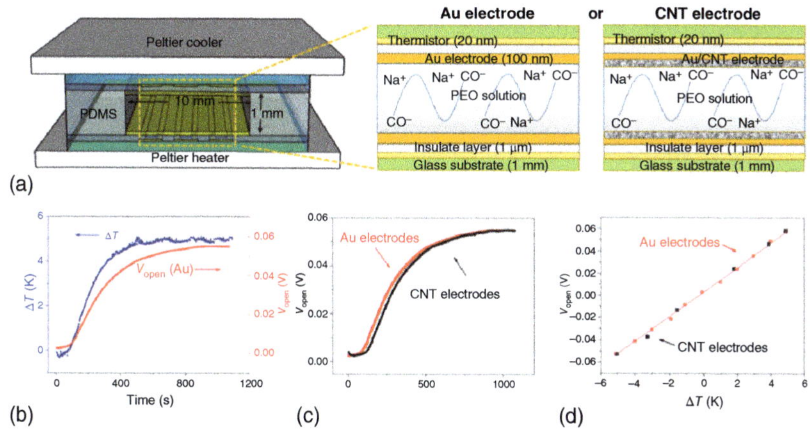

Figure 7.9 Schematic diagram of the ionic thermoelectric supercapacitor device and thermoelectric properties of PEO–NaOH solution. Source: Zhao et al. [7]/Royal Society of Chemistry. (a) Sketch of the ionic thermoelectric supercapacitor device. (b) Measured thermovoltage and heat difference during heating using the Au electrode. (c) The measured thermovoltage during heating using either Au or CNT electrodes. (d) thermovoltag at heat difference values using Au electrodes (red open squares) and CNT electrodes (black solid squares).

for intermittent heat sources such as the sun. At the same time, they discussed the potential improvement of the ITESC concept and predicted that the thermoelectric charge conversion efficiency could be improved if a good thermoelectric electrolyte with low specific heat capacity, low thermal conductivity, large Seebeck coefficient, and high ionic conductivity could be designed.

NaCl–PMSC ionogel electrolyte and CNT–PAM electrodes are used for stretchable i-TE supercapacitors with excellent thermoelectric conversion and electrochemical storage performance [20]. The main body of the NaCl–PMSC ion gel electrolyte is the combination of physically cross-linked CMC–Na and chemically cross-linked PAM to construct a PMSC ion gel electrolyte with cation selectivity. The thermoelectric conversion performance of the ion gel electrolyte was further improved by adding NaCl in the PMSC ion gel, and the maximum Seebeck coefficient could reach 17.1 mV K^{-1}. The elongation at break of the NaCl–PMSC electrolyte is 235% due to the continuous network of the elastic polymer. The stretchable electrode has excellent capacitance characteristics. When the scan rate is 50 mV s^{-1}, the specific surface area capacitance is 38.1 mF cm^{-2}, and the maximum window voltage is 2 V. High-performance i-TE supercapacitors can not only directly convert low-grade waste by converting heat into electricity, but also store the converted electricity in CNT–PAM electrodes. Even after 7500 charge-discharge cycles, the capacity retention of the i-TE supercapacitor can still be maintained at ~86.1%. This unique i-TE supercapacitor simultaneously guarantees stretchability, thermoelectric conversion, and electrical energy storage and has great potential in wearable energy applications.

7.5 Challenges and Opportunities

Thermoelectric ionogels provide an inexpensive, continuous, and sustainable method to generate electricity from low-level heat, which is suitable for multiple application scenarios and maintains safety. There are still many challenges and opportunities for the future development and application of thermoelectric ionogels.

Currently, the challenges of ionogels are the low electron transport and low conversion efficiency in ion gels and the transport efficiency between electrodes and ionogels. Compared with conventional thermoelectric ion transport devices, the electronic conductivity of devices with thermoelectric ion gels is limited because of their low ion mobility. Furthermore, although gel-state electrolytes have the advantages of higher thermoelectric potential and easier encapsulation, the ionic conductivity is lower than that of liquid electrolytes due to steric effects. Thermal diffusion ion gels have high thermoelectric potentials but cannot generate continuous current due to the lack of redox reactions between electrodes and ionophores. In addition to conventional platinum electrodes, nanostructured carbon materials and redox-active materials have also been gradually developed for thermoelectric conversion devices due to their advantages of low cost, high specific surface area, and multifunctionality. Unlike planar materials, the morphology of ionic interfaces between these materials and gel-state electrolytes is worthy of investigation for improving the interface thermoelectric conversion efficiency.

In addition to being used as an electrolyte, thermoelectric ion gels can also be used as electrode materials for thermoelectric devices. Therefore, the study of ionogel thermoelectric devices in terms of electrodes and gel electrolytes is a key challenge for the production of highly efficient integrated hybrid mechanism thermoelectric materials. The formulations used in solid-state thermoelectric batteries and thermocurrent batteries do not apply to thermally diffusive ion gels, including energy conversion efficiencies. Therefore, the conversion formula for thermoelectric devices needs to be further optimized.

In addition to being used in flexible wearable devices, thermoelectric devices with ion gels have also been successfully used in 3C devices, transportation devices and multiple application scenarios after their performance has been improved.

References

1 Zhou, Z., Wan, Y., Zi, J. et al. (2023). *Mater. Today Sustain.* 21: 100293.
2 Shi, X.L., Zou, J., and Chen, Z.G. (2020). *Chem. Rev.* 120: 7399.
3 Massetti, M., Jiao, F., Ferguson, A.J. et al. (2021). *Chem. Rev.* 121: 12465.
4 Cheng, H., He, X., Fan, Z., and Ouyang, J. (2019). *Adv. Energy Mater.* 9: 1901085.
5 Zhao, D., Martinelli, A., Willfahrt, A. et al. (2019). *Nat. Commun.* 10: 1093.
6 Evans, P.N., Boyd, J.A., Leu, A.O. et al. (2019). *Nat. Rev. Microbiol.* 17: 219.
7 Zhao, D., Wang, H., Khan, Z.U. et al. (2016). *Energy Environ. Sci.* 9: 1450.
8 Liu, W., Qian, X., Han, C.-G. et al. (2021). *Appl. Phys. Lett.* 118: 020501.

9 Hu, R., Cola, B.A., Haram, N. et al. (2010). *Nano Lett.* 10: 838.

10 Dubal, D.P., Chodankar, N.R., Kim, D.H., and Gomez-Romero, P. (2018). *Chem. Soc. Rev.* 47: 2065.

11 Yang, P., Liu, K., Chen, Q. et al. (2016). *Angew Chem. Int. Ed. Engl.* 55: 12050.

12 Taheri, A., MacFarlane, D.R., Pozo-Gonzalo, C., and Pringle, J.M. (2019). *Electrochim. Acta* 297: 669.

13 Zinovyeva, V., Nakamae, S., Bonetti, M., and Roger, M. (2014). *ChemElectroChem* 1: 426.

14 Wu, J., Black, J.J., and Aldous, L. (2017). *Electrochim. Acta* 225: 482.

15 Han, C.-G., Qian, X., Li, Q. et al. (2020). *Science* 368: 1091.

16 Dai, H., Zhang, G., Rawach, D. et al. (2021). *Energy Storage Mater.* 34: 320.

17 Alipoori, S., Mazinani, S., Aboutalebi, S.H., and Sharif, F. (2020). *J. Energy Storage* 27: 101072.

18 Cheng, X., Pan, J., Zhao, Y. et al. (2018). *Adv. Energy Mater.* 8: 1702184.

19 Liu, Y., Zhang, S., Zhou, Y. et al. (2020). *Adv. Energy Mater.* 10: 2002539.

20 Yang, X., Tian, Y., Wu, B. et al. (2021). *Energy Environ. Mater.* 5: 954.

8

Alkali Metal Thermal Electrochemical Converter

8.1 Introduction

Carbon neutrality refers to the total amount of carbon dioxide or greenhouse gas emissions produced directly or indirectly by a country, enterprise, product, activity, or individual over a specified time period, through the use of low-carbon energy to replace fossil fuels, afforestation, energy conservation, and emission reduction, etc., to offset the carbon dioxide or greenhouse gas emissions that it produces, to realize positive and negative offsets, and to achieve relatively "zero emissions."

Generally speaking, there are two methods to attain carbon neutrality:

(1) Through the carbon offset mechanism, its carbon emissions are matched by carbon emissions reduced elsewhere. For instance, tree planting and the procurement of renewable energy certificates.
(2) Utilization of zero- or low-carbon technologies. The ultimate goal is to use only low-carbon energy instead of fossil fuels.

Ever since the generation of electric power became feasible for commercial use, electricity has been the most crucial source of energy for all of humankind. Now most electricity is currently generated in centralized thermal power plants. To meet the goal, we need to develop distributed power generation, because it can enhance grid reliability, facilitate energy storage, and accommodate carbon-free alternatives.

Combined heat and power systems are a potential solution for distributed energy requirements, as they generate both electricity and thermal energy simultaneously, thermally regenerative electrochemical systems. The alkali metal thermal-to-electric converter, also known as AMTEC, is one specific type of thermally regenerative electrochemical system. The AMTEC, originally called the sodium heat engine was invented by Joseph T. Kummer and Neill Weber at Ford in 1966. An AMTEC can use either sodium (Na) or potassium (K) as the working fluid, β''-alumina ($Na_2O \cdot MgO \cdot 5Al_2O_3$) can serve as the solid electrolyte (Figure 8.1) [1]. The functioning principle of an AMTEC is depicted in Figure 8.2a [2]. The β''-alumina is one of the most important parts of AMTEC, β''-alumina consists of a layered structure of γ-phase Al_2O_3 blocks separated by monolayer thick ionic conduction planes populated with Na_2O, and it is stabilized with MgO. The ionic conductivity of the β''-alumina is significantly higher than the

Low-Grade Heat Harvesting: Materials, Devices, and Technologies, First Edition. Xiaogang Zhang.

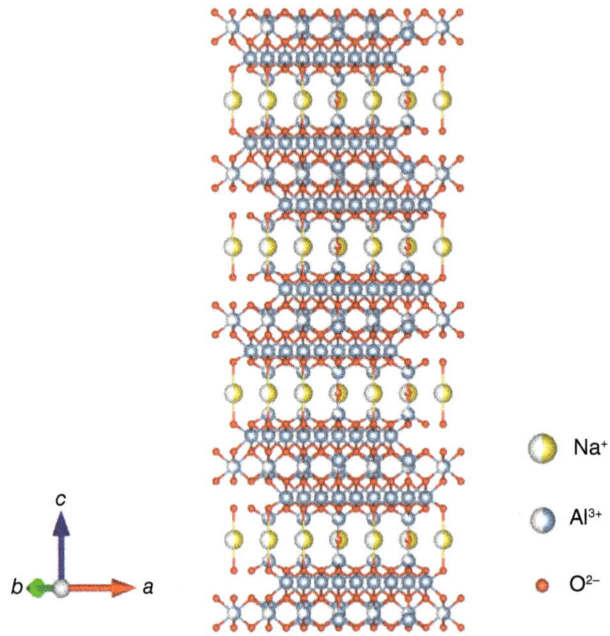

Figure 8.1 Crystal structure of β″-alumina.

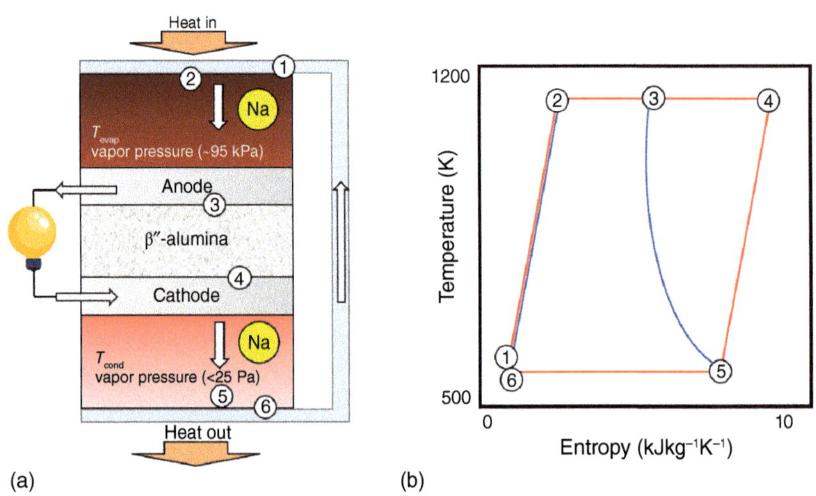

Figure 8.2 (a) Schematic illustration of a single-stage Na-TEC. (b) Temperature–entropy (T–S) diagram of the ideal dual-stage Na-TEC thermodynamic cycle (red lines) overlaid on top of the sodium saturation curves (blue lines).

electronic conductivity. The β″-alumina separates the AMTEC into two distinct regions: the hot area that is under high pressure (20–100 kPa) and high temperature (900–1300 K), and the cold area that is under low pressure (<100 Pa) and low temperature (400–700 K). The hot region is filled with sodium liquid or vapor under high pressure. The low-pressure side is covered by a porous electrode consisting of tiny size granules, which generally range from 0.1 to 0.3 mm.

To describe more clearly, take sodium thermal electrochemical converter as an example, in single-stage operation, liquid sodium first moves into the high-temperature evaporator. Sodium vapor moves through a porous anode across the β″-alumina because of the thermodynamic potential. As a result of the pressure differential between the high-temperature evaporator and the lower-temperature condenser, sodium ions are able to diffuse across the β″-alumina and arrive at the cathode. Simultaneously, the electrons go through the external load, which results in the production of electricity, the electrons eventually arrive at the cathode surface, where they recombine with the sodium ions at the interface between the β″-alumina and the cathode. After leaving the porous electrode, the neutral sodium travels through the vapor space until it reaches the condenser surface. Once there, it gives up the latent heat of condensation and transforms into a liquid state. The resultant liquid metal is returned to the high-pressure cavity by capillary action or an electromagnetic pump. There, it is transformed into high-pressure vapor by the heat that is provided by an external source on the hot side to complete the thermodynamic cycle.

In Figure 8.2b, the ideal thermodynamic cycle can be modeled using six internally reversible processes: (1)–(2) isobaric liquid heating, (2)–(3) vaporization, (3)–(4) isothermal ion expansion through the electrolyte which generates work, (4)–(5) isobaric vapor cooling, (5)–(6) condensation, and (6)–(1) isentropic pumping [3].

The sodium thermal electrochemical converter is an appealing thermally regenerative electrochemical system because it can theoretically attain >90% Carnot efficiency throughout all operational temperature ranges (Figure 8.3). The sodium thermal electrochemical converter has several advantages, including: (i) high operating temperature, resulting in a high theoretical thermal efficiency compared to other thermally regenerative electrochemical systems (up to 45% for a system operating between thermal reservoirs at 1150 and 550 K), (ii) high specific power density (in the range of 0.1 kW/kg_{Na}), (iii) scalability to multiple power outputs (typically in the range of 100 W to 10 kW per module), and (iv) flexibility with heat source (concentrated solar) [4]. While operating at lower temperature ratios, the thermal efficiency of the sodium thermal electrochemical converter competes with various distinct technologies (e.g. single-junction photovoltaics, combustion engines, and thermionics). Traditional power-generating cycles (e.g. Rankine) are built to function at far bigger scales, and so outperform it. The sodium thermal electrochemical converter's specific power and power density can outperform many solid-state generators such as photovoltaics, thermoelectric, and other electrochemical technologies (e.g. fuel cells).

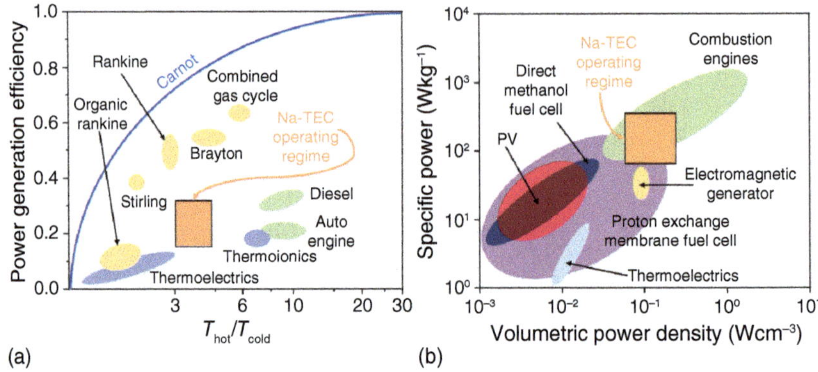

Figure 8.3 (a) Power generation efficiency vs. temperature ratio (between hot and cold reservoirs) for various technologies. Na-TEC efficiency range assumes the use of a dual-stage device. (b) Specific power vs. volumetric power density for various technologies.

8.2 The Single- and Dual-Stage Alkali Metal Thermal Electrochemical Converter

8.2.1 The Single-Stage Sodium Thermal Electrochemical Converter

In 1968, researchers at Ford Scientific Laboratories in Dearborn, Michigan, conceived the AMTEC technology when they discovered and patented the sodium heat engine converter. In the 1980s, liquid anode, single-tube converters with sodium working fluid were devised for AMTECs [5]. In these converters, the liquid sodium served as both the reaction's sodium source and anode. In this instance, the sodium was recirculated using an electromagnetic device. This cell was commonly known as a "liquid anode cell." However, liquid anode converters are low-voltage, high-current devices with a limited power density per unit volume. Connecting the β″-alumina tubes in series is one method for increasing the terminal voltage of the AMTEC [6].

Similar to a concentration cell, the sodium thermal electrochemical converter generates voltage through the oxidation of sodium vapor and the reduction of sodium ions at varying pressures. V_{oc} (open circuit voltage) is determined by the Nernst equation, in which the equilibrium constant is approximated by a pressure ratio, while $V_{a,oc}$ and $V_{c,oc}$ are measured at the electrode/electrolyte interfaces of the anode and cathode, respectively (Eq. 8.1) [7]. As current is drawn, the total voltage V (Eq. 8.2) is the open-circuit voltage reduced by the combined kinetic and mass transfer polarizations (i.e. overpotential) and by the Ohmic potential drop across the electrolyte, which is calculated using the Thevenin equivalent area-specific Ohmic resistance r_{Th} (i.e. ionic resistance of the electrolyte, sheet resistance, and contact resistance in the electrodes, Ω cm²). Both the current density j and the area-specific resistance r_{Th} are normalized to the electrolyte's surface area as follows:

$$V_{oc} = V_{c,oc} - V_{a,oc} \approx \frac{RT_{evap}}{F} \ln \frac{P_a}{P_{coc}} \tag{8.1}$$

$$V = V_{oc} + (\xi_c - \xi_a) - jr_{Th} \tag{8.2}$$

The overpotential is computed using a modified Butler–Volmer equation (Eq. 8.3) [8] that includes both the kinetic and mass transfer polarizations, with a transfer coefficient of 1/2. Ohmic potential drop is dominated by the ionic resistivity of β″-alumina ($1/\sigma_\beta$), which is a function of the crystalline structure, the doping concentration of additives such as MgO or Li_2O, and the operating temperature. Without MgO, oxygen anions (O^{2-}) populate interstitial positions in the β″-alumina conduction plane, reducing the mobility of sodium cations. This electrolyte's high ionic conductivity, which makes it suitable for the sodium thermal electrochemical converter, is the result of the Frenkel transport mechanism, whereby sodium cations in the conduction plane move to vacancy positions left after stable ions transfer to interstitial positions. The model created by Steinbruck (Eq. 8.4) [9] is commonly used to estimate the temperature-dependent ionic conductivity of β″-alumina.

$$j_i = j_{o,i}\left(e^{-\left(\frac{1}{2}\right)\frac{F}{\overline{R}T_{\text{evap}}}\xi_i} - \frac{P_i}{P_{oc,i}}e^{\left(\frac{1}{2}\right)\frac{F}{\overline{R}T_{\text{evap}}}\xi_i}\right) \tag{8.3}$$

$$\sigma_\beta = \left[(1.62\times10^{-5})T_{\exp}\left(\frac{-45.5}{T}\right) + (1.55\times10^{-7})T_{\exp}\left(\frac{3722}{T}\right)\right]^{-1} \tag{8.4}$$

where i in Eq. (8.3) refers to both the anode (a) and cathode (c) and $j_c = j_a$ under charge conservation. The exchange current density j_0 is dependent on temperature, sodium pressure, the electrode material, and its catalytic activity. Due to the rapid kinetics of sodium oxidation relative to its reduction in this device (i.e. $j_{0,a} > j_{0,c}$), the cathode overpotential magnitude is substantially greater than the anode overpotential magnitude. This variable was tabulated for a number of electrode materials. At the electrode/electrolyte interface, higher values of B correspond to more rapid kinetics. In the limit of $B \to \infty$, there is no charge transfer kinetics resistance at the electrode/electrolyte interface, and only mass transfer resistance contributes to the overpotential [10].

$$j_{0,i} = B\sqrt{\frac{p_{oc,i}p_{sat}(T)}{T}} \tag{8.5}$$

At low pressures, sodium transport from the electrode/electrolyte interface to the bulk plenum is controlled by molecular effusion through the electrodes' microscopic pores. Due to the long mean free path of sodium at the saturation pressure of typical condenser temperatures $P_{sat}(T_{cond})$, transport at the cathode occurs under the free molecular flow regime (Knudsen number $K_n \geq 10$). The saturation pressure of sodium vapor, $P_{sat}(T)$, is calculated using an empirical Antoine relation. In this range of K_n, the rate at which sodium molecules enter and depart the apertures of a porous structure is proportional to $P\sqrt{T}$; this phenomenon is known as thermal transpiration. At open circuit, there is no net mass transfer between the cathode/electrolyte interface and the condenser, so Eq. (8.6) can be used to calculate the open-circuit pressure P_{oc} at the cathode/electrolyte interface. In the anode, the mean free path is much shorter than the pore dimension, so the pressure at the anode/electrolyte interface is equivalent to the saturation pressure at the evaporator temperature [4].

$$\frac{P_{c,oc}}{\sqrt{T_{evap}}} = \frac{P_{sat}(T_{cond})}{\sqrt{T_{cond}}} \Rightarrow P_{c,oc} = P_{sat}(T_{cond})\sqrt{\frac{T_{evap}}{T_{cond}}} \tag{8.6}$$

P_i, the pressure at the electrode/electrolyte interface, is a function of the current density in a closed circuit (see Eq. 8.3). In the free molecular flow regime ($K_n \geq 10$), the gas kinetic theory is used to calculate the sodium flow pressure decrease. Equation (8.7) describes the pressure decrease at the cathode as sodium departs the electrode/electrolyte interface, assuming an accommodation coefficient of one. Similarly, as sodium traverses the gas–liquid interface in the condenser, the pressure decreases (Eq. 8.8) [11].

$$\Delta P = \frac{j}{\varphi F} \tag{8.7}$$

$$\Delta P = \frac{j}{F}\sqrt{2\pi\bar{R}MT_{cond}} \tag{8.8}$$

The electrode pores can be modeled as cylindrical channels with an average pore diameter of d_p. For a tube with a small aspect ratio ($L_e/d_p <\approx 1$), the pressure drop ΔP_K for a rarefied gas transporting through a porous electrode at temperature T_{evap} is given by Eq. (8.9), where R_K is the mass transfer resistance for Knudsen flow. Finding the ratio of molecules that flow completely through the channel to the total number of molecules that penetrate the channel boundary yields this equation. The dimensionless morphology factor G is typically employed for modeling purposes as an empirical variable, and it is a function of the electrode's thickness, porosity, and pore diameter. On the anode side of the device, the sodium vapor is at a significantly higher pressure, and the flow is continuous ($K_n \rightarrow 0$). Using the Hagen–Poiseuille law (Eq. 8.10), the pressure drop through a cylindrical pore is calculated, where R_v is the mass transfer resistance for viscous flow, $\mu(T)$ is the temperature-dependent sodium vapor viscosity, and $\mu(T)$ is the temperature-dependent sodium vapor density. A significant difference in absolute pressures (4–6 orders of magnitude) necessitates the calculation of the anode pressure drop, even though it is negligible in comparison to the cathode pressure drop. Particularly required for determining the limiting current [12].

$$\Delta P_K = \frac{3G}{8\pi}\frac{j}{F}\sqrt{2\pi\bar{R}MT_{evap}} = R_K\left(\frac{jM}{\varphi F}\right) \rightarrow G = \frac{8\pi}{3\varphi}\left(1 + \frac{L_e}{d_p}\right) \tag{8.9}$$

$$\Delta P_v = \frac{32\mu L}{d_p^2\rho\varphi}\left(\frac{jM}{F}\right) = R_v\left(\frac{jM}{\varphi F}\right) \tag{8.10}$$

With precise knowledge of the pressures at the electrode/electrolyte interfaces, it is possible to calculate the limiting current. The limit of mass transfer occurs when kinetic losses are negligible (i.e. $B \rightarrow \infty$). Equation (8.11) describes the mass transfer-limited voltage in a single-stage sodium thermal electrochemical converter when Ohmic losses are also disregarded. The limiting current represents the utmost reaction rate of the system, such that no additional voltage change can increase the

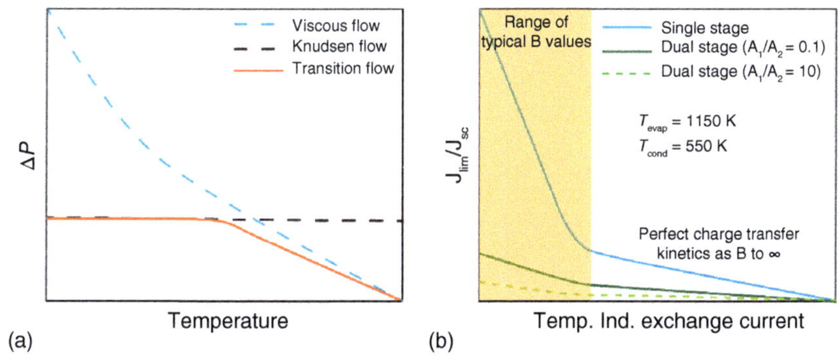

Figure 8.4 (a) Plot of the sodium vapor flow pressure drop in the transitional regime vs. temperature ($L_e = 1\ \mu m$, $\varphi = 90\%$, $G = 10$, $j = 0.5\ A\ cm^{-2}$). In the limit of $Kn \to 0$, the vapor transport is described with a viscous flow model whereas in the limit of $Kn \to \infty$, the vapor transport is described with a Knudsen flow model. (b) Ratio of the limiting current and the short circuit current vs. the temperature-independent exchange current.

current. Consequently, the limiting current is found mathematically in the limit where $V \to -\infty$ (i.e. an infinitely negative bias does not result in an increase in the current). In Eq. (8.11), this occurs if either the denominator (cathode pressure) or the numerator (anode pressure) reaches infinity or zero, respectively. Only the latter case is physically significant, so Eq. (8.12) [13] describes the limiting current. Calculations reveal that the limiting current density is significantly greater than the short-circuit current density (i.e. $j_{lim}/j_{sc} > 1$; see Figure 8.4b), indicating that mass transfer in the anode does not limit the cycle. Rather, the cathode charge transfer, Ohmic losses, and cathodic mass transfer all have a significant impact on the power density [14].

$$\lim_{B\to\infty} V = \frac{\overline{R}T_{evap}}{F} \ln\left(\frac{P_{a,oc} - \Delta P_{slip}(\rho(T_{evap}), \mu(T_{evap}))}{P_{c,oc} + \left(\frac{1}{\varphi} + \frac{3G}{8\pi}\right)\frac{j}{F}\sqrt{2\pi\overline{R}MT_{evap}} + \frac{j}{F}\sqrt{2\pi\overline{R}MT_{cond}}}\right) \tag{8.11}$$

$$P_{a2,oc} = \Delta P_{slip}(\rho(T_{evap}), \mu(T_{evap}), j_{lim}) \to j_{lim} \approx \frac{d_p^2\rho\varphi}{32\mu L_e}\left(\frac{F}{M}\right)P_{a,oc} \tag{8.12}$$

The power density ($jV = j^2 r_{load}$) depends on the current density and the area-specific external load resistance r_{load}, which is normalized to the total electrolyte area (e.g. $r_{load} = AR_{load}$). To calculate the utmost power, the power density is implicitly differentiated with respect to r_{load}. This operation results in the load-matching condition (Eq. 8.13), where r_{ct} represents the charge transfer resistance. The charge transfer resistance is experimentally measured by obtaining the difference between the high and low-frequency real-axis intercepts of a Nyquist plot from an electrochemical impedance spectroscopy measurement. This load-matching condition is nonlinear with the current density, but it can be numerically solved to determine the utmost power [15].

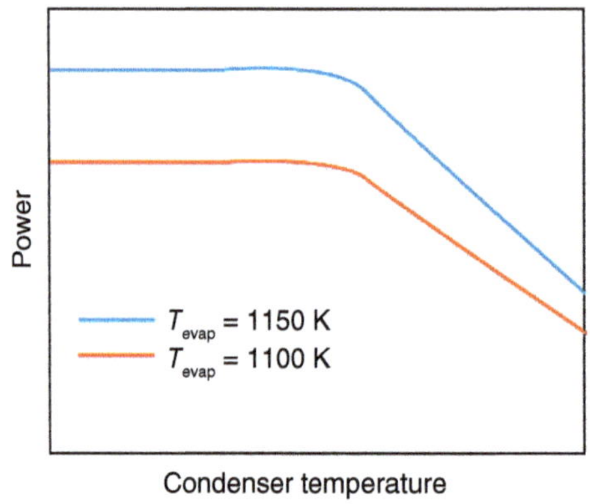

Figure 8.5 Maximum power of a single-stage Na-thermal electrochemical converter (Na-TEC) as a function of condenser temperature ($B = 120\,\mathrm{A\,K^{1/2}\,N^{-1}}$, $G = 50$, $r_{Th} = 0.25\,\Omega\,\mathrm{cm}^2$).

$$r_{load} = r_{Th} + (r_{ct,c} - r_{ct,a}) \rightarrow r_{ct,i} = -\frac{\partial \xi_i}{\partial j} \tag{8.13}$$

The utmost power as a function of condenser temperature is depicted in Figure 8.5. As the temperature of the condenser decreases, the cathode pressure falls and the open-circuit voltage rises. At lower temperatures, this higher open-circuit voltage is counterbalanced by an increase in the total overpotential. Consequently, the maximal power density saturates at low condenser temperatures and exhibits virtually constant power over a broad range of condenser temperatures. In fact, the utmost power difference between the coldest possible condenser temperature where sodium solidifies (T 370 K) and 575 K in Figure 8.5 is only 1.5%.

Thermal efficiency is the ratio between output electricity and input heat. The thermal parasitic loss is the sum of (i) heat input that does not contribute to power generation but merely bypasses the thermo-electrochemical process and is transferred to the condenser, and (ii) heat that is lost directly to the environment from the device due to inadequate insulation. To analyze the ideal thermodynamic cycle, only heat that bypasses the thermo-electrochemical process is taken into account in the parasitic loss term. This energy loss manifests itself as conduction heat transfer through the device's walls and capillary wick, as well as radiative heat transfer from the cathode to the condenser. According to energy conservation for the cycle, the heat input equals the aggregate of heat outputs and generated power. Parasitic loss is affected by operating conditions, materials, geometry, and environment. For parametric calculations, the ratio of parasitic heat loss to total heat input is used. The electrolyte area is used to normalize both terms. The efficiency is subsequently rewritten as Eq. (8.14) [16].

$$\begin{aligned}\eta &= \frac{jV}{(1+f/(1-f))\left(jV + \frac{jM}{F}\left(h_{fg}(T_{cond}) + \int_{T_{cond}}^{T_{evap}} c_p(T)\mathrm{d}T\right)\right)} \\ &= \frac{V(1-f)}{V + \frac{M}{F}\left(h_{fg}(T_{cond}) + \int_{T_{cond}}^{T_{evap}} c_p(T)\mathrm{d}T\right)}\end{aligned} \tag{8.14}$$

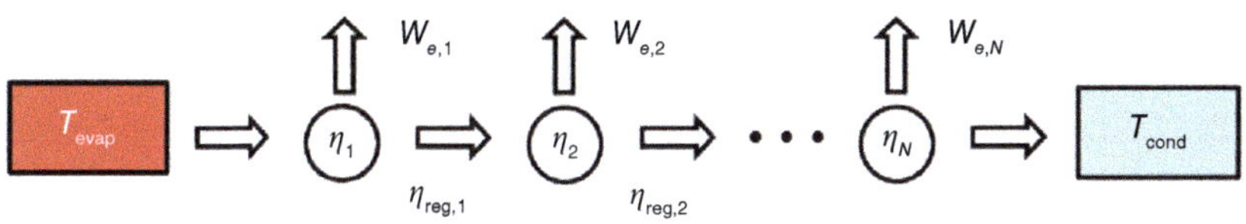

Figure 8.6 Schematic showing N heat engines interacting between two thermal reservoirs with regeneration efficiencies $\eta_{reg,i}$ of the heat output from the first $N-1$ engines, each producing power $W_{e,i}$.

In order to increase the efficacy of a single-stage sodium thermal electrochemical converter, the isothermal expansion phase of the cycle can be divided into two stages, one at the evaporator temperature and the other at an intermediate temperature. This efficiency improvement is supported by the following thermodynamic argument. Consider first, as depicted in Figure 8.6, N heat engines interacting between two thermal reservoirs and exchanging heat. Assume there are $N-1$ regenerators between the heat engines that regenerate the rejected heat, with each regenerator's efficiency η_{reg} being variable. The total efficacy of these N engines with $N-1$ regenerators in between them, interacting between two thermal reservoirs with a constant temperature, is represented by Eq. (8.15). This expression is derived by calculating the ratio of total work to heat input $(W_{e,1}+W_{e,2}+\cdots+W_{e,N})/(1/W_{e,1})$ for a limited number of engines (e.g. $N \leq 4$) until the recursive pattern is recognized.

Using this framework, it can be presumed that each engine is a single-stage sodium thermal electrochemical converter with a 90% Carnot efficiency. When the conversion step in the sodium thermal electrochemical converter is broken down into multiple stages, each stage can be depicted as a single-stage sodium thermal electrochemical converter. Using Eq. (8.15), the total efficiency of a multistage sodium thermal electrochemical converter operating between reservoirs with fixed temperatures can be determined for any number of stages.

$$\eta_N = \eta_1 + \sum_{i=1}^{N-1}\left[(\eta_{i+1})\prod_{j=1}^{i}[\eta_{\mathrm{reg},j}(1-\eta_j)]\right] \tag{8.15}$$

8.2.2 The Dual-Stage Sodium Thermal Electrochemical Converter

The dual-stage sodium thermal electrochemical converter regarded as a series of engines is more efficient than the most efficient existing single sodium thermal electrochemical converter, with the former able to increase the maximum efficiency and power output density by up to 17.5% and 40.6%, respectively, over the latter. In addition, the maximum efficiency of the dual-stage sodium thermal electrochemical converter can be improved by 34.1% and 24.8%, respectively, compared to that of the dual-stage sodium thermal electrochemical converters. The objectives of this part are twofold: (i) to improve the model of a dual-stage sodium thermal electrochemical converter, and (ii) to investigate the optimal performance β′′-alumina solid electrolyte on the design and main losses key parameters: the β′′-alumina solid electrolyte thickness, the current density, and the temperature intermediate between the two stages.

Two singular AMTEC devices are coupled according to continuity and energy conservation constraints so that the second AMTEC device operates using heat supplied by the first. Consideration is given to heat transfer by conduction and radiation, as well as pressure decreases through the porous electrodes, for both processes, β″-alumina solid electrolyte on the appropriate geometrical parameters. The additional supposition of the intermediate temperature for both phases is β″-alumina solid electrolyte on Limia et al.'s efforts to minimize losses due to overpotential and Ohmic contributions. Such an intermediate temperature may play a significant role in the performance regimens of the individual and collective behaviors.

Concentrate on the analysis of the impact of certain key parameters (such as extreme temperatures, current densities, cross-sectional areas, and thickness β″-alumina solid electrolyte) on the chosen merit: the power output density and thermal efficiency of the overall device. Thus, special emphasis is placed on the function of the intermediate temperature, as it is discovered that the merit only reaches their respective maximum values within a relatively limited range of values.

As depicted in Figure 8.7, the operational principles of a dual-stage sodium thermal electrochemical converter are similar to those of a single-stage sodium thermal electrochemical converter. At temperature T_H, q_H represents the heat transfer from the heat source to the evaporator plenum in the first stage of the sodium thermal

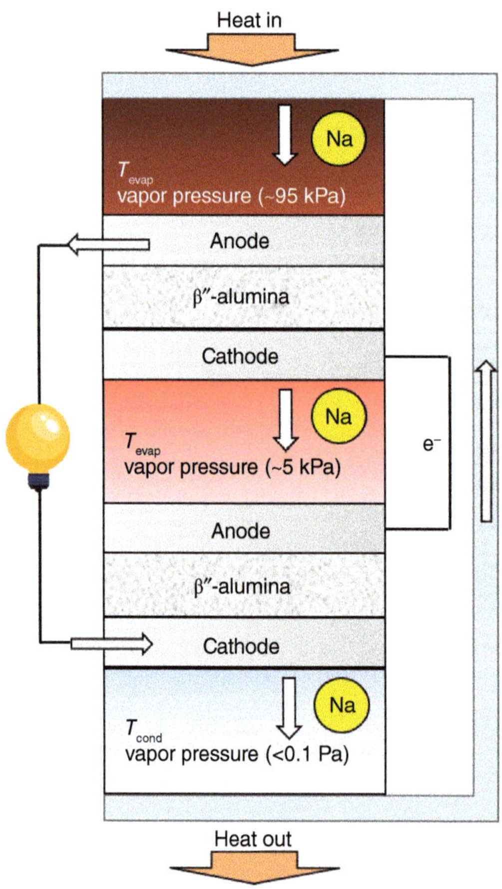

Figure 8.7 The schematic diagram of a dual-stage sodium thermal electrochemical converter. Source: Adapted from Peng et al. [17].

electrochemical converter. q_C represents the heat flux from the condenser plenum at temperature T_C of the second stage sodium thermal electrochemical converter to the surrounding environment at temperature T_0. The temperature of the plenum's intermediate temperature is tin. $A_{1,2}$ is the surface area of the first/second stage electrolyte, $D_{1,2}$ is the thickness of the first/second stage electrolyte, and $P_{1,2}$ is the power output of the first/second stage sodium thermal electrochemical converter. $P = P_1 + P_2$ represents the total power output of the dual-stage sodium thermal electrochemical converter device. In the initial stage of operation, liquid sodium is heated and vaporized in the high-temperature plenum. Sodium cations expand isothermally through the β″-alumina solid electrolyte from the high-temperature plenum into the intermediate plenum, where the pressure is equal to the vapor pressure at an open circuit plus the pressure drop caused by the recombination of electrons and ions, before flowing through the porous electrode at the close circuit. The sodium vapor dissipates the heat q_{in} and cools at constant pressure, which is completely regenerated in the subsequent stage of the sodium thermal electrochemical converter. The sodium expands isothermally from the intermediate plenum to the low-temperature condenser plenum, where it cools and condenses to give off q_C of waste heat.

The thermodynamic operation of the first stage sodium thermal electrochemical converter is similar to the single sodium thermal electrochemical converter. Thus, the first stage thermal first-law-based efficiency, defined as the ratio of the total power output to total heat input, η_1 is given by [17]:

$$\begin{aligned} \eta_1 &= \frac{P_1}{q_H} \\ &= \frac{J_1 A_1 \left[\frac{RT_H}{F} \ln[1 + p_{sat}(T_H)\mu D_1/(RT_H)] - V_{ac,1} - V_{R,1}\right]}{\begin{array}{c} J_1 A_1 \left[\frac{RT_H}{F} \ln[1 + p_{sat}(T_H)\mu D_1/(RT_H)] - V_{ac,1} - V_{R,1}\right] \\ + J_1 A_1 ML/F + J_1 A_1 c_P (T_H - T_{in}) M/F + A_1 \sigma \left(T_H^4 - T_{in}^4\right)/z \end{array}} \end{aligned} \tag{8.16}$$

where

$$\begin{aligned} V_{R,1} = J_1 D_1 & \left[1.62 \times 10^{-5} T_H \exp(-45.5/T_H) + 1.55 \right. \\ & \left. \times 10^{-7} T_H \exp(3722/T_H)\right] \end{aligned} \tag{8.17}$$

$$\begin{aligned} V_{ac,1} = & -\frac{2RT_H}{F} \ln\left[\frac{1}{2}\left[\frac{J_1^2 T_H}{B_1^2 p_{sat}^2(T_H)} + 4\right]^{\frac{1}{2}} - \frac{1}{2J_1 p_{sat}(T_H)}\right] \\ & + \frac{2RT_H}{F} \ln\left[\frac{1}{2}\left[\frac{J_1^2 \sqrt{T_H T_{in}}}{B_1^2 p_{sat}(T_H) p_{sat}(T_{in})} + 4\left[1 + \frac{\left(1 + \frac{3G_1}{8\pi}\right)\sqrt{2\pi MRT_H T_{in}}\frac{J_1}{F}}{p_{sat}(T_{in})\sqrt{T_H}}\right]\right]^{\frac{1}{2}}\right. \\ & \left. + \frac{1}{2}\frac{J_1 (T_H T_{in})^{\frac{1}{4}}}{B_1 \sqrt{p_{sat}(T_H) p_{sat}(T_{in})}}\right] \end{aligned} \tag{8.18}$$

$P_{sat}(T) = 109.678 - 5383.2/T$ is the saturation vapor pressure at temperature T, $V_{R,1}$ is the first stage ionic β″-alumina solid electrolyte voltage, $V_{ac,1}$ is the first stage over potential difference, G_1 is the first stage pressure loss geometric factor, B_1 denotes the first stage charge-exchange coefficient, J_1 is the first stage current density, L and M are the sodium vaporization latent heat and molecular weight, c_P is the molar specific heat, z is the radiation reduction factor, F is the Faraday constant, σ the Stefan–Boltzmann constant, μ is a coefficient, and R is the gas constant. The specific relational expression among T_H, μ, D_1, and T_{in} should satisfy the following relation [18]:

$$\frac{p_{sat}(T_H)}{1 + p_{sat}(T_H)\mu D_1/(RT_H)} = p_{sat}(T_{in})\sqrt{\frac{T_H}{T_{in}}} \tag{8.19}$$

As shown in Figure 8.7, the heat q_{in} from the first stage is fully utilized by the second-stage sodium thermal electrochemical converter. The thermal efficiency η_2 can be written as:

$$\eta_2 = \frac{P_2}{q_{in}} = \frac{J_2A_2\left[\frac{RT_{in}}{F}\ln[1 + p_{sat}(T_{in})\mu D_2/(RT_{in})] - V_{ac,2} - V_{R,2}\right]}{J_1A_1ML/F + J_1A_1c_P(T_H - T_{in})M/F + A_1\sigma\left(T_H^4 - T_{in}^4\right)/z} \tag{8.20}$$

where in this case

$$V_{R,2} = J_2D_2\left[1.62\times10^{-5}T_{in}\exp(-45.5/T_{in}) + 1.55 \times 10^{-7}T_{in}\exp(3722/T_{in})\right] \tag{8.21}$$

$$\begin{aligned} V_{ac,2} = &-\frac{2RT_{in}}{F}\ln\left[\frac{1}{2}\left[\frac{J_2^2T_{in}}{B_2^2p_{sat}^2(T_{in})}+4\right]^{\frac{1}{2}} - \frac{1}{2}\frac{J_2\sqrt{T_{in}}}{B_2p_{sat}(T_{in})}\right] \\ &+\frac{2RT_{in}}{F}\ln\left[\frac{1}{2}\left[\frac{J_2{}^2\sqrt{T_{in}T_C}}{B_2^2p_{sat}(T_{in})p_{sat}(T_C)}+4\left[1+\frac{\left(1+\frac{3G_2}{8\pi}\right)\sqrt{2\pi MRT_{in}T_C}\frac{J_2}{F}}{p_{sat}(T_C)\sqrt{T_{in}}}\right]\right]^{\frac{1}{2}}\right. \\ &\left.+\frac{(J_2T_{in}T_c)^{\frac{1}{4}}}{2B_2\sqrt{p_{sat}(T_{in})p_{int}(T_C)}}\right] \end{aligned} \tag{8.22}$$

$V_{R,2}$ indicates the second stage ionic β″-alumina solid electrolyte voltage, $V_{ac,2}$ is the second stage over potential difference, G_2 means the second stage pressure loss geometric factor, J_2 indicates the second stage current density, and B_2 stands for the second stage charge-exchange coefficient. Similarly, some parameters including T_{in}, μ, D_2, and T_C should satisfy the following equation:

$$\frac{p_{sat}(T_{in})}{1 + p_{sat}(T_{in})\mu D_2/(RT_{in})} = p_{sat}(T_C)\sqrt{\frac{T_{in}}{T_C}} \tag{8.23}$$

the thermal efficiency, η, of the dual-stage system can be expressed as:

$$\eta = \frac{P}{q_H} = \eta_1 + \eta_2(1 - \eta_1) \tag{8.24}$$

where the power output of the whole device is,

$$
\begin{aligned}
P = P_1 + P_2 &= q_H - J_2A_2ML/F - J_2A_2c_P(T_{\text{in}} - T_{\text{C}})M/F - A_2\sigma\left(T_{\text{in}}^4 - T_{\text{C}}^4\right)/z \\
&= J_1A_1\left[\frac{RT_{\text{H}}}{F}\ln[1 + p_{\text{sat}}(T_{\text{H}})\mu D_1/(RT_{\text{H}})] - V_{\text{ac},1} - V_{R,1}\right] \\
&\quad + J_2A_2\left[\frac{RT_{\text{in}}}{F}\ln[1 + p_{\text{sat}}(T_{\text{in}})\mu D_2/(RT_{\text{in}})] - V_{\text{ac},2} - V_{R,2}\right]
\end{aligned}
\tag{8.25}
$$

A thermodynamic comparison between the single-stage and dual-stage sodium thermal electrochemical converter can be conducted in two distinct ways: (i) with a fixed hot-side temperature and a fixed cold-side temperature, and (ii) with a fixed hot-side temperature and a fixed heat input. When dealing with actual prototypes, it is customary to compare devices using the latter method, in which the total heat input is fixed and both devices are assumed to be in contact with the same hot-side thermal reservoir. Incorporating reheating and regeneration from the first-stage isobaric cooling, these boundary conditions permit the dual-stage condenser temperature to be lower than its single-stage counterpart. This lower dual-stage condenser temperature is determined by establishing the total heat input for a given intermediate temperature and a single-stage condenser temperature. Using this method, the temperature of the dual-stage condenser must be recalculated whenever the operating power changes, as the total heat input also varies. This recalculation is unnecessary because the heat input in the limit of zero current is the most conservative case. In this limit, it is sufficient to solve for the temperature of the dual-stage condenser as a function of the temperature of the single-stage condenser by equating the heat input for both systems.

In the absence of a prototype, approach (i) is more adequate for a theoretical thermodynamic analysis. To compare the thermodynamic operation of the single-stage and dual-stage sodium thermal electrochemical converter, it is necessary to consider a scenario in which both devices are in contact with the same thermal reservoirs so that the evaporator and condenser temperatures are identical for each device. In addition, it is supposed that the regeneration efficiency between the first and second stages is one hundred percent. This assumption is legitimate when heat loss to the environment is disregarded in the parasitic loss term. This assumption is further supported by the fact that a minimal quantity of heat will flee the intermediate plenum via the reheat path into the second-stage electrolyte. In addition, electrode materials whose performance would otherwise degrade rapidly at elevated evaporator temperatures can be utilized in the second stage, where their superior performance at lower temperatures can be exploited. Porous molybdenum is an excellent choice for the second-stage electrode for this purpose.

Area circumscribed by the temperature–entropy (T–S) diagram is proportional to the efficiency of a reversible thermodynamic cycle. In the reversible limit, the dual-stage cycle is inferior to the single-stage cycle because a portion of the heat input occurs at an intermediate temperature, reducing the total area of the dual-stage cycle's T–S diagram. In addition, the total voltage generated by the dual-stage sodium thermal electrochemical converter is lower than that generated by the single-stage converter due to significant overpotential contributions from three electrodes (the first-stage cathode, the second-stage anode, and the second-stage cathode) rather than just one (the cathode). Adding a second stage increases the magnitude of the Ohmic resistance as well. The dual-stage sodium thermal electrochemical converter becomes competitive only when the parasitic loss inherent in realistic devices, which is the dominant factor influencing system efficiency, is accounted for. In general, since the average operating temperature decreases as the number of stages increases, one can anticipate that the parasitic loss will be reduced in the dual-stage device due to its lower average operating temperature. According to previous experimental findings, the highest-performing single-stage prototypes have operated at optimum efficiency with a parasitic loss fraction of 45%. Using this value for the parasitic loss fraction, a regime map can be generated to aid in the selection of a single-stage or dual-stage sodium thermal electrochemical converter. This map depicts the most desirable system based on power density and dual-stage parasitic loss fraction. When the dual-stage sodium thermal electrochemical converter has a parasitic loss greater than 45% ($f_{dual} > 45\%$), a single-stage architecture is always more efficient. So long as the power density is below the regime boundary, it is more efficient to operate with a dual-stage sodium thermal electrochemical converter if the parasitic loss fraction is less than 45%. The regime boundary is determined by fitting a polynomial to the intersection sites of the efficiency vs. power density curves of the single- and dual-stage sodium thermal electrochemical converters at different f_{dual} values.

Due to conduction loss to the condenser (via the device enclosure walls and capillary wick) and radiation loss (via direct radiative energy exchange) between the cathode and the condenser, previous single-stage sodium thermal electrochemical converter prototypes have struggled to achieve a high level of efficiency. The average hot-side temperature is lower in the dual-stage device, which reduces the total parasitic conduction loss. In addition, the second stage will radiate directly to the condenser at T_{int} as opposed to T_{evap}, reducing the heat that bypasses the thermo-electrochemical process without the need for additional radiation shielding.

For a general combination of operating variables, it is possible to determine the temperature that produces the highest power and the highest efficiency. The maximal power density is a function of the intermediate temperature that increases monotonically. In the limit where the dual-stage sodium thermal electrochemical converter resembles a single-stage sodium thermal electrochemical converter, the maximal power density occurs. Therefore, for high-power applications, it is necessary to maximize the intermediate temperature. Maximum efficacy relative to the intermediate temperature is dependent on j_1 and A_1/A_2.

A more efficient alternative to the conventional single-stage sodium thermal electrochemical converter has been developed: the dual-stage sodium thermal electrochemical converter. The thermodynamic operating space for this technology is investigated using expressions for the voltage, thermal efficiency, and maximal power density of a dual-stage sodium thermal electrochemical converter. Dual-stage sodium thermal electrochemical converters are limited by overpotential and Ohmic contributions from two stages, whereas single-stage sodium thermal electrochemical converters are only affected by a single electrochemical stage. Consequently, the single-stage sodium thermal electrochemical converter obtains a higher power density with the same total electrolyte area. However, the dual-stage sodium thermal electrochemical converter has a greater thermal efficacy. Using a simple heat transfer model, it has been demonstrated that the conduction and radiative parasitic heat losses through the device are reduced with a dual-stage configuration because the average device temperature is lower, resulting in a smaller temperature difference across which these heat transfer mechanisms operate. This manifests as a decrease in the fraction f of parasitic heat loss (i.e. the fraction of heat input lost parasitically). In addition, the dual-stage device supports regeneration and reheating procedures, allowing for more efficient utilization of the supplied heat. Assuming that high-performance electrodes ($R_{hx}W$ for the first stage and Mo for the second stage) are utilized in a device, an application regime map was developed to aid in the selection of a single- or dual-stage sodium thermal electrochemical converter based on the power density and the parasitic loss fraction. In applications where a high-power density is required (e.g. because β″-alumina is scarce), a single-stage device should be utilized owing to its prospective thermal management advantage, a dual-stage sodium thermal electrochemical converter should be used when a high thermal efficiency is required (e.g. owing to high fuel costs). A simple heat transfer model predicts that the practical efficiency of a dual-stage sodium thermal electrochemical converter can approach 26.5%, which is 7.5% points greater than the highest efficiency single-stage sodium thermal electrochemical converter, whose maximum practical efficiency is 19%, while the maximum power density approaches 2.22 kW m^{-2}.

8.3 The Alkali Metal Thermal Electrochemical Converter Devices

8.3.1 Working Fluids for AMTEC

Comprehensive modeling of AMTEC cells is an appropriate method for supporting designs and evaluating the performance of AMTEC cells. When predicting the performance of an AMTEC, several primary operational factors must be evaluated. These aspects are interrelated and can be addressed by four submodels, namely, the flow model, the thermodynamic model, the electrochemical model, and the electric circuit model. Due to the intricacy of the cell geometry and the fact that the vapor flow is in the transition or free-molecular regime, only a small number of researchers have attempted to establish a flow model to calculate vapor pressure losses. Schock et al. [19] and Hendricks et al. [20] created intricate thermodynamic, electrical, and vapor flow models for vapor anode multi-tube AMTEC cells. Tournier et al. described a lumped analytical model of liquid anode single-tube and vapor anode multi-tube AMTEC cells. Experimental data and model results for Mo, NbN, and TiN electrodes agreed well. The Institute for Space and Nuclear Power Studies at the University of New Mexico. (UNM-ISNPS) devised the AMTEC Performance and Evaluation Analysis Model (APEAM). This integrated cell model is comprised of four submodels: a sodium vapor pressure loss model, a radiation/conduction heat transfer model, a cell electrochemical model, and a two-dimensional cell electric circuit model. The Air Force Research Laboratory/Phillips Laboratory (AFRL/PL) has effectively validated the completely integrated cell model (APEAM) using experimental data of single- and multi-tube cells ground-tested in a vacuum. The accompanying descriptions pertain to multi-tube vapor anode Na-AMTEC cells. It should be noted, however, that the model can also be applied to multi-tube vapor anode K-AMTEC cells. This modeling is applicable to vapor-fed, liquid anode, multi-tube Na-, or K-AMTEC cells with a few modifications [21].

Historically, lithium (Li), sodium (Na), and potassium (K) were deemed possible alkali metal working fluids for AMTECs. Lithium was eliminated from consideration for the numerous reasons including the prices, reserves, and melting point. Both sodium and potassium are appropriate AMTEC working fluids. The current experience with sodium as working fluids for AMTECs is extensive, whereas the data on potassium use is limited. Using potassium working fluid and potassium-β''-alumina solid electrolyte is an alternative method for developing AMTECs with high efficiency that has not been seriously considered until now. In reality, the physical properties of potassium and sodium differ significantly. It is difficult to reach a conclusion and quantify the advantages of potassium converters based merely on the relative comparison of the physical properties of the potassium and sodium working fluids [22]. Consequently, a systematic analysis is required. Few studies on K-AMTECs have been published in the scientific literature. Williams et al. [23] noted that K-AMTECs with Mo electrodes demonstrated improved electric power output and conversion efficiency compared to Na-AMTECs. Huang et al. [24] suggested that a potassium converter could provide better performance than a sodium converter if the evaporator wick in the former was designed with smaller pores to increase the capillary pressure head. Barkan et al. [25] investigated the efficacy of a nonrecirculating potassium mini-converter at temperatures between 820 and 1073 °C. The experimental results obtained in the liquid anode mode of operation demonstrated that the potassium converter produced the same amount of electrical power as the finest sodium mini-converter evaluated, but at 50 K lower β''-alumina solid electrolyte temperature. El-Genk and Tournier [26] utilized the APEAM to investigate and compare the performance and operating limits of Mo-41% Re sodium and potassium vapor anode, multi-tube AMTECs of identical design. At the same load voltage, the potassium converter with the remote evaporator wick was able to deliver the same amount of electrical power as the sodium converter while operating at 50 K lower hot side temperature. When operated at the same load voltage ($V_L < 2.8$ V) and hot side temperature ($T_{hot} < 1150$ K), the potassium converter with the remote evaporator wick can provide an average of 1.2 W more power than the reference sodium converter with 2% higher conversion efficiency.

8.3.2 Electrode for AMTEC

The electrode is one of the most significant components of AMTECs, contributing 15–19% of the total power output degradation. Consequently, the search for high-performance electrodes is a viable strategy for enhancing the power output and efficacy of AMTECs. This should involve materials, geometric optimization, and a suitable vacuum level on the cathode side of the AMTEC. Desirable electrode properties include [27] (1) Good electrical conductivity; (2) Good physical bonding with the β″-alumina solid electrolyte membrane; (3) A thermal expansion coefficient similar to that of β″-alumina solid electrolyte; (4) High permeability to sodium vapor; (5) High corrosion resistance to sodium vapor; (6) Low material loss rate by chemical reaction or sublimation; and (7) Slow grain growth and material migration. None of the refractory electrode materials, including TiB_2, LaB_6, TiN, TiC, TiN–TiC, NbN, and NbC, satisfy all of the aforementioned requirements. Among the aforementioned properties, (1)(2)(3)(4)(7) are the most sought-after in the selection of electrode materials, and there are two widely used parameters for estimating the performance of electrode materials. One is the temperature-independent charge-exchange current coefficient B, which measures the contact between the electrode and electrolyte and is related to the electrode's operational longevity via particle size. The other is the dimensionless factor for pressure losses, G_E, which describes the pressure decrease caused by the diffusion of neutral sodium ions at the β″-alumina solid electrolyte–cathode interface of the porous electrode. High-performance electrodes should have a greater value of B, indicating fewer internal charge-exchange polarization losses, and a lower value of G_E, indicating fewer concentration losses. In addition, the current collector/electrode contact resistance R_{cont} is utilized to assess the performance of electrodes.

The evolution of electrode materials can be roughly divided into four stages, beginning with the earliest metal material (Mo), followed by ceramic electrodes (TiN, TiC, TiN–TiC, LaB_6, TiB_2, etc.), refractory alloy electrodes ($R_{hx}W$), and the most recent mixed-conducting (ionic and electronic) metal-oxide electrodes (Mo/Na_x-TiO_2 and TiN/Na_x-TiO_2). Using refractory metal coatings of a few micrometers in thickness and deposited by chemical vapor deposition or sputtering, these values can be met extraordinarily well. Additionally, it was suggested that the vacuum level on the low-temperature side of AMTECs should be maintained below 20 Pa to reduce polarization and concentration losses. If this is not possible and the value must be greater than the critical value, it may be maintained as near as practicable to the critical value.

8.4 Challenges and Opportunities

The sodium thermal electrochemical converter can have a transformative effect on micro combined heat and power (micro-CHP) systems ranging from 1 to 10 kW as the need for distributed power rises in tandem with the global residential energy demand. Before these facilities can be utilized in the real world, numerous issues must be resolved.

What are the thermodynamic constraints on a dual-stage sodium thermal electrochemical converter? In order to determine the optimum power density and thermal efficiency, thermodynamic, electrochemical, and heat transfer variables are considered. Temperatures at the evaporator, intermediate plenum, and condenser determine the operational limits of this technology. The voltage is determined by the sodium vapor pressure established by these temperatures, which influences the electrochemical polarizations at each stage. The optimum power density is then determined by optimizing the operating variables independently. Due to the irreversibility resulting from the use of two electrochemical stages, the dual-stage sodium thermal electrochemical converter is less suitable for high-power applications than its single-stage counterpart. However, the thermal efficiency is most sensitive to thermal parasitic heat losses in both single-stage and dual-stage devices. Due to the lower average temperature for bypass heat transfer, these losses may be reduced in a dual-stage sodium thermal electrochemical converter.

To what extent can a reduced-order model be used to design a dual-stage sodium thermal electrochemical converter so that it operates with the highest practicable efficiency? In conjunction with the thermodynamic model, a reduced-order modeling approach can be used to analyze the thermal performance of a dual-stage sodium thermal electrochemical converter early in the design process, guide the selection of materials, and determine the total system size with minimal computational resources and acceptable accuracy. The reduced-order model includes effective thermal conductivities to account for complex corrugated geometries as well as apparent surface emissivity to account for the effect of radiation shields.

To construct a full-scale prototype of a dual-stage sodium thermal electrochemical converter in order to determine whether its thermal efficiency can surpass that of previous single-stage prototypes ($\eta \leq 19\%$). The unavailability of a reliable high-temperature metal-to-ceramic barrier for β″-alumina disks is one of the technical obstacles impeding operation at higher temperatures. In each stage of this module, a thermocompression bond based on aluminum was used to create a seal, but its effectiveness was limited to temperatures below 650 °C. Air brazing with Ag-8CuO brazing alloy has shown promise at higher temperatures, but further experimentation is required before a prototype can be created. This module must be optimized in future design iterations to reduce thermal parasitic losses. Additionally, a liquid return path must be integrated into the design in order to construct a closed-loop prototype. The thermal efficacy of this closed-loop prototype can then be compared to that of single-stage devices previously constructed. In future research, the reduced-order thermal model can also be refined. Importantly, the effect of sodium pressure decreases from various design components must be considered. For instance, thermal efficiency can be optimized based on the length of the corrugated structure in the condenser, given that a longer corrugated path reduces parasitic heat conduction but increases the sodium vapor pressure drop. None of the component pressure reductions have been accounted for in the reduced-order model as of yet.

To construct a low-temperature capillary pump with a non-wetting porous structure, additional effort is required. Recommendation: a second law analysis of the capillary pumping mechanism. This experiment's entropy generation can be calculated using techniques similar to those applied to heat pipes, where the irreversibility between the evaporator and the upper flange is caused by heat transfer through finite temperature differences, diffusion of sodium vapor, and frictional losses during fluid transport. Through a parametric survey of temperature, pressure, pore size, porosity, and local mean curvature, an optimization study can be conducted to minimize entropy generation across the condensation interface in microchannels by employing the same analyses used for evaporation interfaces. Once thermodynamically favorable conditions have been established, additional capillary pumping experiments can be conducted to expand upon the limited results presented in this paper. These investigations should be conducted at higher pressures in order to complete a pump curve, which defines the pumping limits effectively. In conjunction with an appropriate system curve, a pump curve can be used to design a pump for a sodium thermal electrochemical converter with two stages. Future experiments must also continue to investigate the distinction between thermally conductive and thermally insulating porous inserts. Finding a more resilient electrically insulating material to mount the electrodes on the upper flange is a practical experimental enhancement. Although the high-temperature cement has performed adequately, its deterioration increases the uncertainty of the voltage signals used to verify the sodium flow. Additionally, the conjugate transport model used to guide these investigations can be enhanced. For instance, radiation within the porous structures can be incorporated into the model in order to determine a more precise temperature distribution within the experimental coupon. Additionally, a parametric study can be used to optimize the diameter and height of the experimental coupon while allowing for convection on its outer surface. Lastly, if a high-temperature vacuum barrier for SS316 can be created, this experiment can be conducted under a vacuum. While differing vacuum pressures are enforced in the evaporator plenum, the upper plenum would be left exposed to the atmosphere. In the rarefied and transitional gas regimes, sodium vapor would be limited by viscous transport and natural convection rather than binary diffusion. Finally, other materials may be considered for the proposed capillary pump, and the breakthrough pressure experiment may be used to determine the pumping limits of novel coupons. Between pressure increments, an extremely conservative waiting period can be used to reduce experimental uncertainty. In addition, to prevent sodium leakage between the porous coupon and the SS316 cylinder, a micro-weld can be applied to the edge between these two components. By directly sintering metal particles inside the cylinder, an even tighter boundary closure can be achieved. The temperature-dependent contact angle of liquid sodium can also be measured independently using a sessile drop experiment. This information can be used to rectify the experiment's large breakthrough pressure variation. The breakthrough pressure experiment can be supplemented with other experimental techniques, such as measuring the saturation of the structure with a porosimeter, to reduce the uncertainty in the permeability measurements.

References

1 Kennedy, J.H. (1977). The β-aluminas. In: *Solid Electrolytes*, vol. 105 (ed. S. Geller), 105–141. Berlin Heidelberg: Springer.
2 Weber, N. (1974). *Energy Conv.* 14: 1.
3 Moran, M.J. and Tsatsaronis, G. (2017). Engineering thermodynamics, 1. In: *CRC Handbook of Thermal Engineering*, 2e (ed. R.P. Chhabr), 1–112. Boca Raton: CRC Press.
4 Wu, S.Y., Xiao, L., and Cao, Y.D. (2009). *Int. J. Energy Res.* 33: 868.
5 Bankston, C., Cole, T., Jones, R., and Ewell, R. (1983). *J. Energy* 7: 442.
6 Cole, T. (1983). *Science* 221: 915.
7 Bard, A.J. and Faulkner, L.R. (2001). *Electrochem. Methods* 2: 580.
8 Koryta, J., Dvorak, W., and Kavan, L. (1987). *Principles of Electrochemistry, Second Edition.* Chichester: John Wiley & Sons.
9 Steinbruck, M. (1993). Investigations of beta-alumina solid electrolytes for application in AMTEC cells. In: *Proceedings of 28th Intersociety Energy Conversion Engineering Conference*, p. 799. American Chemical Society, Atlanta, GA.
10 de los Rios Ramos, N.D., Onea, A., Scherrer, S. et al. (2015). Direct energy conversion of heat to electricity using AMTEC, In: *2015 5th International Youth Conference on Energy (IYCE)*, p. 1. Institute of Electrical and Electronics Engineers, Pisa, Italy.
11 Tournier, J.-M. and El-Genk, M.S. (1999). *J. Thermophys. Heat Transfer* 13: 117.
12 Kennard, E.H. (1938). *Kinetic Theory of Gases*. New York: McGraw-Hill.
13 Schuller, M., Azimov, U., and Lalk, T. (2002). *J. Electrochem. Soc.* 149: A1432.
14 Underwood, M.L., Williams, R.M., Ryan, M.A. et al. (1992). An AMTEC vapor-vapor, series connected cell. In: *AIP Conference Proceedings*, p. 1331. AIP Publishing, Albuquerque, New Mexico.
15 Paul, R., Etacheri, V., Wang, Y., and Lin, C.-T. (2019). Carbon Based Nanomaterials for Advanced Thermal and Electrochemical Energy Storage and Conversion. *Elsevier.*
16 Tournier, J.-M., El-Genk, M.S., Schuller, M., and Hausgen, P. (1997). An analytical model for liquid-anode and vapor-anode AMTEC converters. In: *AIP Conference Proceedings*, p. 1543. AIP Publishing. Albuquerque, New Mexico.
17 Peng, W., Zhang, X., Ye, Z., and Chen, J. (2018). *Energy Conv. Manag.* 168: 230.
18 Wu, S.-Y., Xiao, L., Cao, Y., and Li, Y.-R. (2010). *Appl. Energy* 87: 452.
19 Schock, A., Noravian, H., Or, C., and Kumar, V. (1997). Parametric analyses of AMTEC multitube cells and recommendation for revised cell design. AIP Conference Proceedings, p. 1395. AIP Publishing, Albuquerque, New Mexico.
20 Hendricks, T.J., Borkowski, C.A., Huang, C. (1998). Development & experimental validation of a SINDA/FLUINT thermal/fluid/electrical model of a multi-tube AMTEC cell. In: AIP Conference Proceedings, p. 1491. AIP Publishing, Albuquerque, New Mexico.
21 (a) Tournier, J.-M. and El-Genk, M.S. (1998). Sodium vapor flow regimes and pressure losses on cathode side of multitube AMTEC cell. In: AIP Conference

Proceedings, p. 1595. AIP Publishing, Albuquerque, New Mexico; (b) Tournier, J.M. and El-Genk, M. (1999). *J. Appl. Electrochem.* *29*: 1263.

22 Foust, O. (1972). *Sodium-NaK Engineering Handbook. Volume I. Sodium Chemistry and Physical Properties*. Gordon and Breach, Science Publishers, Inc.

23 Williams, R., Kisor, A., Ryan, M. et al. (1994). Potassium beta-double-prime-alumina/potassium/molybdenum electrochemical cells. In: *Intersociety Energy Conversion Engineering Conference*, p. 3833. American Institute of Aeronautics and Astronautics, Monterey, CA.

24 Huang, C., Hendricks, T.J., and Hunt, T.K. (1998). *Performance Analysis of a Potassium-BASE AMTEC Cell*. Ann Arbor, MI (US): Advanced Modular Power Systems, Inc.

25 Barkan, A., Hunt, T., and Thomas, B. (1999). *Potassium AMTEC Cell Performance*. SAE Technical Paper, 1999-01-2702.

26 El-Genk, M.S. and Tournier, J.-M. (1931). *Energy Conv. Manag.* 2002: 43.

27 Williams, R., Loveland, M., Jeffries-Nakamura, B. et al. (1990). *J. Electrochem. Soc.* 137: 1709.

9

Thermally Regenerative Electrochemical Cycle and Other Techniques

9.1 Introduction

Pollution and global warming are becoming more and more prominent nowadays due to the use of fossil fuels. Harmful emissions such as CO_2, SO_x, and NO_x and particles less than 2.5 μm in diameter have led to serious disasters for humans and nature. Clean and sustainable power generation technologies can replace the current coal/oil method and its associated problems. There is also a large amount of low-temperature heat in various forms (e.g. below 130 °C), such as solar thermal, geothermal, and waste heat from vehicles and industry. In addition, a significant amount of primary energy generated worldwide is discharged as low-grade heat due to inefficient energy conversion. Therefore, reusing and converting this low-temperature heat is a promising solution to address global energy problems. Converting this thermal energy into other types of energy, such as electricity, remains a major challenge.

Among the various methods of heat generation, thermoelectric technology is one of the most promising and intensively studied. Thermoelectric devices are based on the development of a voltage across a conducting material under a temperature gradient, known as the Seebeck effect. This effect is described by the Seebeck coefficient, which is defined as the ratio between the voltage difference and the temperature difference ($S = \mathrm{d}V/\mathrm{d}T$). The efficiency of thermoelectric devices is determined by the Figure of merit ($\mathrm{ZT} = S^2\sigma T/k$), where σ and k are the electrical and thermal conductivities, respectively [1, 2]. For practical cryogenic applications of thermoelectrics, a ZT in the range of 3–4 is desirable [1, 3]. However, despite considerable progress in recent years, the ZT remains below 1.5 at temperatures <150 °C. A major reason for this is that S is only about 0.2 mV K^{-1} for modern materials. In electrochemical systems, there is a phenomenon similar to the Seebeck effect, which is due to the temperature dependence of the electrode potential in electrochemical reactions. For an electrochemical reaction ($\mathrm{A} + ne^- \rightarrow \mathrm{B}$), the temperature coefficient (α) of the electrode potential is $\alpha = \frac{\mathrm{d}V}{\mathrm{d}T} = \frac{S_\mathrm{B}-S_\mathrm{A}}{nF}$, where S_A and S_B are the partial molar entropies of A and B, respectively, and F is Faraday's constant. This coefficient for electrochemical systems (α) is much larger (about an order of magnitude) than the Seebeck coefficient in thermoelectric systems, which is the main advantage of electrochemical systems over thermoelectric systems [4].

Low-Grade Heat Harvesting: Materials, Devices, and Technologies, First Edition. Xiaogang Zhang.

There are two ways to use such high-temperature coefficient electrochemical systems. The first approach is to use a pyrogen battery with the same configuration as a thermoelectric device, called a thermoelectric battery or liquid thermoelectric [5]. The second approach is to create an electrochemical cycle called a thermally regenerative electrochemical cycle (TREC). In TREC, an electrochemical cell is constructed with two different electrodes whose temperature coefficient (α_{cell}) is the difference between the temperature coefficients of the positive ($\alpha+$) and negative ($\alpha-$) electrodes negative electrode (−, T1)/electrolyte (T1)/positive electrode (+, T1).

The temperature coefficient A of the battery core is the difference between the temperature coefficient of the positive electrode α_+ and the negative electrode α_-:

$$\alpha_+ = \left(\frac{dE_+}{dT}\right)_{\text{ios}} \tag{9.1}$$

$$\alpha_- = \left(\frac{dE_-}{dT}\right)_{\text{ios}} \tag{9.2}$$

$$\alpha_{\text{cell}} = \left(\frac{dV_{\text{oc}}}{dT}\right)_{\text{ios}} = \left(\frac{dE_{\mp}}{dT}\right)_{\text{ios}} - \left(\frac{dE_-}{dT}\right)_{\text{ios}} = \frac{\Delta S_{\text{cell}}}{nF} \tag{9.3}$$

where E is the electrochemical potential of the electrode reaction, V_{oc} is the open-circuit voltage of the whole battery in isothermal mode, and ΔS_{cell} is the total entropy change in the whole-cell reaction. Temperature coefficient of the whole cell (α_{cell}) can be positive or negative, depending on the temperature coefficient of the positive and negative electrodes. When $\alpha_{\text{cell}} > 0$, the battery is first heated to a high temperature (T_{H}) and discharged at T_{H}. Then, the battery is cooled to a low temperature (T_{L}) with the circuit open, and the open-circuit voltage becomes small due to the positive α_{cell}, the battery is then charged to the initial state of charge (SOC) at T_{L} to complete a cycle (Figure 9.1a). If $\alpha_{\text{cell}} < 0$, the cell is first cooled to T_{L} and discharged, then it is heated to T_{H} and charged to the initial state SOC (Figure 9.1b). In both cases, the charging curve is lower than the discharging curve, and network is generated by this voltage difference. The cell voltage depends on temperature, so a thermodynamic cycle can be created by discharging the battery at T1 and charging it at T2. If the charging voltage at T2 is lower than the discharging voltage at T1, the net energy produced by this voltage difference is due to heat absorbed at higher temperatures, similar to a heat engine with an upper bound on Carnot efficiency [6].

The charge curves for both thermodynamic cycles are lower than the discharge curves in comparison to usual charge–discharge processes; hence, network is created during both cycles due to the voltage difference. TREC drew a lot of interest in the 1980s; however, the emphasis was mostly on high-temperature thermal energy. Their study focuses on extracting high-temperature heat (>500 °C). However, low-grade heat harvesting (e.g. >130 °C) using TREC has received little attention because of low efficiency due to voltage gain, which is analogous to resistive losses. As a result, developing high-efficiency TRECs for harvesting low-grade heat is critical.

TREC is a thermodynamic cycle in which heat is absorbed at T_{H} and rejected partly at T_{L}. The difference is used to generate electrical power. The temperature–entropy

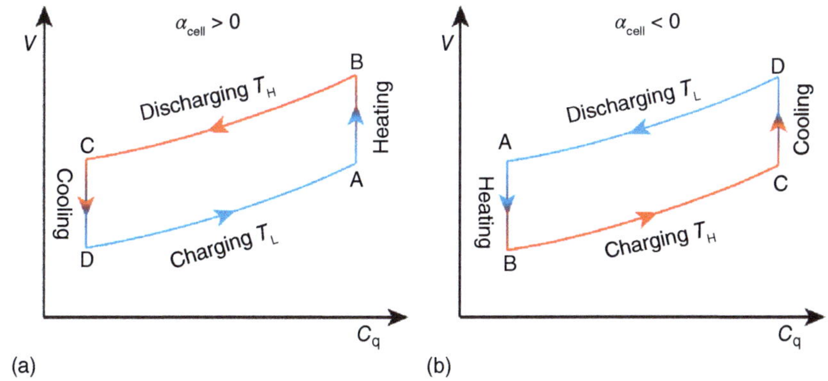

Figure 9.1 Voltage-charge capacity diagram of TREC for (a) $\alpha_{cell} > 0$ and (b) $\alpha_{cell} < 0$.

plot (Figure 9.2a) clearly demonstrates this. TREC efficiency (η) can be estimated by dividing network (W) by thermal energy input. When $\Delta T = (T_H - T_L)$ is not great and the enthalpy change ΔH and entropy change ΔS are considered to be the same at T_H and T_L, the maximum W is $\Delta T \Delta S$. The energy input necessary to complete the cycle is divided into two parts: the heat absorbed at T_H ($Q_H = T_H \Delta S$) and the external heat required to raise the system temperature (Q_{HR}). Through heat recovery, a portion of the heat discharged from the cooling process can be reused in the heating process. Q_{HR} is expressed as $Q_{HR} = (1 - \eta_{HR}) C_p \Delta T$, where C_p is the total heat capacity of the electrochemical cell and η_{HR} is the efficiency of heat recovery, defined as the ratio of the heat transfer from the cooling process to the heating process to the energy required to heat the battery from T_L to T_H. Therefore, η can be expressed as:

$$\eta = \frac{W}{Q_H + Q_{HR}} = \frac{\Delta T \Delta S - E_{loss}}{T_H \Delta S + (1 - \eta_{HR}) C_p \Delta T} \tag{9.4}$$

where E_{loss} is the energy loss due to the battery resistance ($I(R_H + R_L)$), I is the current used for discharging and charging, and R_H and R_L are the internal resistances at T_H and T_L, respectively. Note that $\Delta T \Delta S = \alpha_{cell} Q_c \Delta T$, where Q_c is the charge capacity of the battery and α_{cell} is the temperature coefficient of the electrochemical cell. Then the efficiency becomes

$$\eta = \eta_c \frac{I - I(R_H + R_L)/|\alpha_{cell}|\Delta T}{1 + \eta_c(1 - \eta_H)/|Y|} \tag{9.5}$$

where $Y = \alpha_{cell} Q_c / C_p$. The efficiency depends on $Y = \alpha_{cell} Q_c / C_p$ when internal resistance is disregarded. The Figure of merit y is similarly defined for a particular electrode material as $\alpha Q_c / C_p$, where Q_c is the specific charge capacity and C_p is the specific heat of the electrode. Therefore, Y is a dimensionless term that describes the need for high-efficiency materials. As a result, it is evident that materials with greater temperature coefficient (α), specific charge capacity (Q_c), and specific heat (C_p) have better thermoelectric conversion efficiency. Efficiency can also be increased via low-pressure polarization and effective heat recovery. Since α, Q_c, and C_p are not connected together, they can be tweaked separately to maximize η.

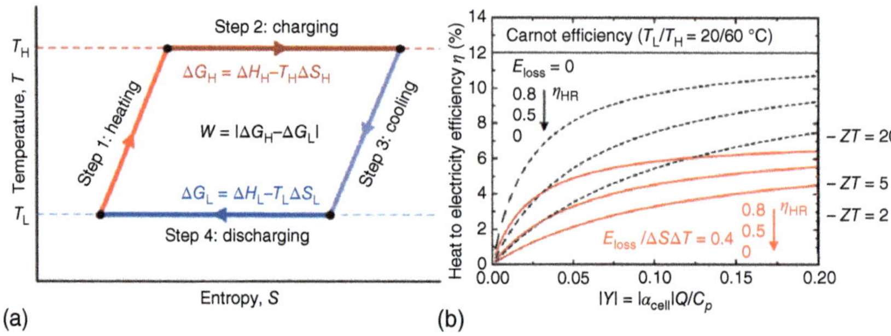

Figure 9.2 (a) Temperature–entropy (T–S) diagram of a thermally regenerative electrochemical cycle (TREC) between T_L and T_H. (b) Thermal to electricity efficiency of TREC when T_L/T_H is 20/60 °C. Black dotted lines correspond to zero energy losses (E_{loss}) from cell resistance. Red lines correspond to a E_{loss} = 40% of the theoretical energy gain ($\Delta T \Delta S$). For each color, the graphs correspond to heat recuperation efficiencies η_{HR} of 0.8, 0.5, and 0 from top to bottom. The ZT values on the right-hand axis are theoretical values required by thermoelectric devices (working between 20 and 60 °C) to reach corresponding efficiencies. Source: Lee et al. [7]/Springer Nature.

The formula for Y appears to be similar to the Figure of merit Z in thermoelectric materials.

The efficiency η vs. Y at T_L/T_H of 20/60 °C is presented in Figure 9.2b to better compare TREC and other technologies. This figure displays a Carnot efficiency (η_c) of 12.0%. When $Y = 0.1$ (e.g. $\alpha_{cell} = 2\,\text{mV}\,\text{K}^{-1}$, specific capacity ($Q_c$) of each electrode is $75\,\text{mAh}\,\text{g}^{-1}$, specific heat ($C_p$) is $2.7\,\text{J}\,\text{g}^{-1}\,\text{K}^{-1}$) and $\eta_{HR} = 50\%$ are met, η can reach 4.5% and 7.5% for $E_{loss}/\Delta T \Delta S$ of 0.4 and 0, 4.5% and 7.5% are achieved. These outcomes highlight TREC's potential for low-grade heat collection.

In addition to conversion efficiency, power density is a crucial factor in the conversion of energy. For the TREC system, the maximum efficiency output is $P_{max} = \frac{(\alpha_{cell}\Delta T)^2}{S(R_H+R_L)}$, when $I = \frac{\alpha_{cell}\Delta T}{2(R_H+R_L)}$, and the specific efficiency is often 1–2 orders of magnitude lower than that of thermoelectric devices, ranging from 1 to $10\,\text{W}\,\text{kg}^{-1}$ in quantity. The battery's high electrical resistance, which results from ions moving through the battery considerably more slowly than electrons, is the major cause of the low power density. How to enhance battery kinetics to boost power density has been a significant area of current study.

9.2 Progresses of TRECs

TREC was initially proposed by Case in 1886, when he described a system based on Sn|Sn(II), Cr(III), Cr(II)|carbon with T_L/T_H of 10/90 °C. Before the 1970s, the major focus was on collecting high-temperature heat (>500 °C) for aerospace and nuclear applications employing molten metal/halide batteries. Hammond et al. demonstrated a TREC system based on the $Fe(CN)_6^{3-}/^{4-}$, $Cu(NH_3)_4^{2+}|Cu(NH_3)^{2+}$ system in 1979, which converted low-grade heat directly from solar energy to electrical energy. As depicted in Figure 9.3, a pump circulates the reaction products

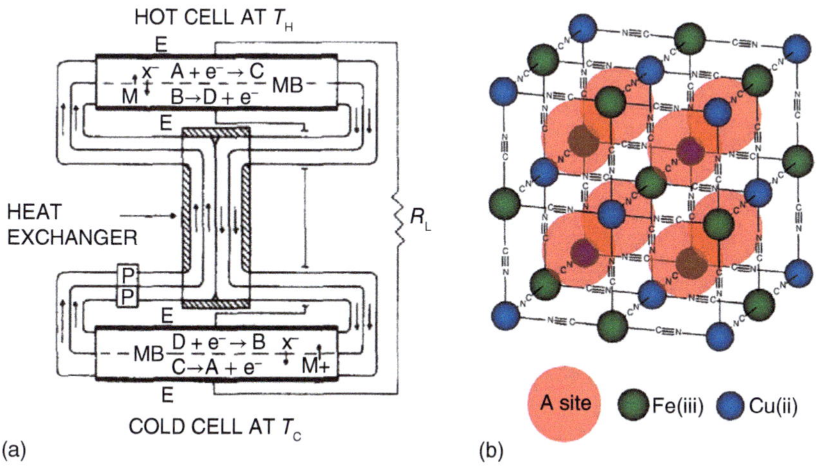

Figure 9.3 (a) Schematic of the $Fe(CN)_6^{3-/4-}$, $Cu(NH_3)_4^{2+}/Cu(NH_3)^{2+}$ system. In this figure, the redox pairs are represented by A/C and B/D in general. The electrodes are marked by E, semipermeable ion transport membranes by MB, pumps by *P*, external load as *L*, and the current-carrying ions as M^+ and X^-. (b) Structure of Prussian blue analog (PBA) copper hexacyanoferrate, A site can be cations (Li, Na, and K) that can intercalate and de-intercalate in and from the lattice. Source: Lee et al. [8]/Springer Nature.

between one cell at T_H (hot state) and another cell at T_L (cold state). The two half-reactions in each cell are divided by a membrane that only permits charge carriers of M^+ and X^- to pass through. The cycle is produced by the continuous electrochemical reaction of A and B in the hot cell and the reaction of C and D in the cold room to create A and B. The voltage differential between the two cells supplies power to an external load (R_L) due to temperature changes. The reaction products from one cell are pumped to the other, resulting in continuous operation and reactant regeneration. Two redox pairings, $Cu(NH_3)_4^{2+}/Cu(NH_3)^{2+}$ (corresponding to A/C) and $Fe(CN)_6^{4-/3-}$ (corresponding to B/D), result in a high α_{cell} of 2.9 mV K^{-1}, which is one of the highest values recorded to date. The specific reactions (reverse reactions) in the hot and cold chambers are provided by:

$$Cu(NH_3)^{2+} + Fe(CN)_6^{4-} \leftrightarrow Cu(NH_3)^{+} + Fe(CN)_6^{3-} + 2NH_3 \tag{9.6}$$

The design is reported to produce 50–75% of η_c between 30 and 90 °C, assuming 100% heat recovery. However, it is constrained by high internal resistance and suffers from poor stability due to $Cu_2Fe(CN)_6$ precipitation within hours of operation.

The limited charge capacity and high specific heat of reaction despite high-temperature coefficient, however, are issues with previously constructed systems. In order to solve this problem, Lee et al. developed the initial Prussian blue analog (PBA)-based TREC approach. High specific charge capacity, low specific heat, and quick kinetics are all characteristics of PBA. In PBA, the core of each cube can include transition metal ions that are connected by cyano groups and produced by alkali ions (like or). PBA also has an open structure, which contributes to its quick kinetics and stable ion insertion and ejection. PBA includes compounds like

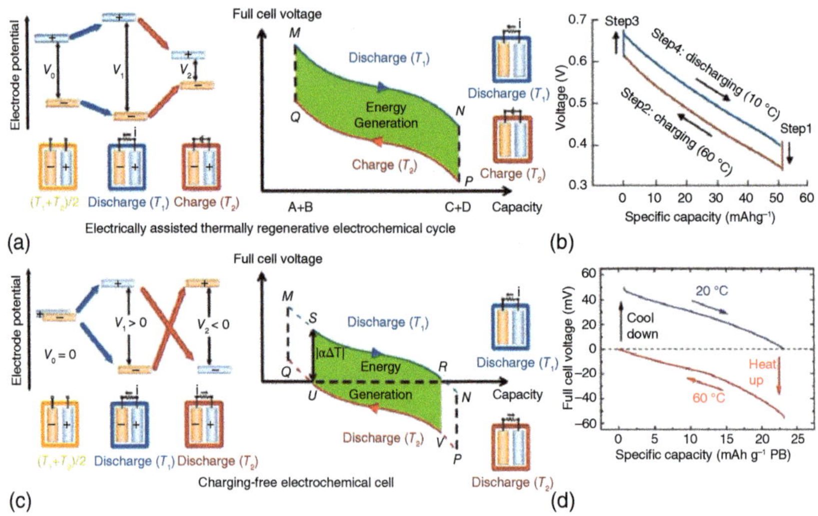

Figure 9.4 Schematics and performances of TREC cells with (a, b) and without (c, d) electrical assistance. (a) Electrode potential changes in positive and negative electrodes and the full cell in an electrically assisted thermal cycle. (b) A voltage plot of the cell vs. the specific capacity of CuHCF for one thermodynamic cycle. (c) Electrode potential changes in positive and negative electrodes and the full cell in a charging-free thermal cycle. (d) Cycling operation of the charging-free system. The cell is operated between 20 and 60 °C [7, 9].

copper hexacyanoferrate (CuHCF), the structure of which is seen in Figure 9.4. Carbon electrodes are enhanced in performance using CuHCF.

The presented TREC system comprises a copper hexacyanoferrate cathode (CuHCF) in the $NaNO_3$ electrolyte and a Cu anode in the aqueous electrolyte $Cu(NO_3)_2$. NO_3^- moves via an ion exchange membrane in this arrangement. The electrode material has low polarization, low specific heat, and a high charge capacity, and the temperature coefficient of the whole cell is as high as 1.2 mV K^{-1}. The battery produced 5.2 J g^{-1} of electric energy when it was charged at T_H (60 °C) and discharged at a higher voltage at T_L (10 °C) (Figure 9.4a,b). Even without heat recovery, the thermoelectric conversion efficiency reaches 3.7%, which is equivalent to 25% of η_c. The conversion efficiency may be further improved to 5.7% with a cautious estimate of η_{HR} of 50%. The shown TREC device needs an additional power source to fully recharge the battery. The charge-free electrochemical cell that the authors of this work created is depicted in Figure 9.4c as a solution to this flaw. The general energy-collecting process is identical to the one in the preceding example. The two electrodes, however, have the same electrochemical potential in the neutral state, and the electrodes' temperature coefficients have the opposite sign. As a result, the battery's polarity changes as the temperature shifts from T1 to T2, and the battery may be drained at either temperature without using an external voltage. Making sure that the voltage difference caused by temperature change is larger than the battery voltage is a crucial component of upgrading the system. The full-cell voltage during thermodynamic cycling is displayed as a function of time in Figure 9.4d. The corresponding η is 0.68% without any heat recovery and 2.0% with η_{HR} of 0.7.

To avoid cathode and anode mixing, Lee et al.'s TREC system calls for a specialized ion-exchange membrane. Such membranes, however, raise the system's total cost and decrease system productivity and long cycle life since it will be difficult to avoid ion permeability after extended operation. Membraneless systems can therefore be employed but only provided the electrolyte-soluble compounds do not result in unfavorable side effects beyond the targeted half-cell processes.

A nickel hexacyanoferrate (NiHCF) cathode and an Ag/AgCl anode were employed in a study by Yang et al., which did not cause any negative side effects, as follows:

$$\text{Cathode: } \mathrm{KNiFe(CN)_6 + K^+ + e^- \rightarrow K_2NiFe(CN)_6} \tag{9.7}$$

$$\text{Anode: } \mathrm{Ag + Cl^- \rightarrow AgCl + e^-} \tag{9.8}$$

The system's temperature coefficient is 0.74 mV K^{-1}. There are no side reactions between the ions in each electrode. Ion-selective membranes are therefore not required and may be swapped out for less costly porous separators. The system displays a low capacity fading rate and strong Coulombic efficiency under long-term operation at various current rates. After 50 cycles at C/2 and an η_{HR} of 0.7, the equivalent thermoelectric efficiency (η) approaches 2.9% (24% of η_c). The increased cost of this system is a result of the more expensive Ag electrodes; however, if an electrode that can substitute Ag can be discovered, the application prospects can be further improved.

The crucial step in developing a good charging-free cell is selecting the electrode materials so that the voltage change brought on by a change in temperature is larger than the voltage of the cell. Additionally, the cell voltage must be close to zero at both T1 and T2's average temperatures in order for the polarity transition of the cell to happen. According to the criteria mentioned above, they chose 0.3 M $K_3Fe(CN)_6$/0.5 M $K_4Fe(CN)_6$ an aqueous solution as a negative electrode and Prussian Blue [$KFe^{II}Fe^{III}$(CN)6, PB] as a positive electrode with 2.5 M KNO_3 as its electrolyte. Two electrodes were separated by a section of Nafion membrane. Positive and negative electrodes have temperature coefficients of 1.46 and 0 mV K^{-1}, respectively. The temperature coefficient for the whole cell was −1.45 mV K^{-1}. Both a cold source temperature of 20 °C and a hot source temperature of 60 °C were used to test the proposed electrochemical cell. According to study findings (Figure 9.5), the system efficiency for converting heat to electricity was around 0.68% without heat regeneration and 2.0% with 70% heat regeneration. In total, 5.67% and 16.67% of Carnot efficiency, respectively, are represented by these efficiencies [9]. It takes a ZT of 0.9 to reach a 2% efficiency compared to a thermoelectric system that runs between the two temperatures described above. The electrochemical system efficiency rises to 3.2% with improved heat regeneration efficiency, or at 85% heat regeneration, in which case the ZT necessary for such efficiency in thermoelectric systems is 2.1. Additionally, the system's performance has been consistent for around 500–1000 cycles.

The temperature coefficient of different PBAs was researched by Gao et al. The purple cobalt hexacyanoferrate (CoHCFe-pp), CuHCFe, and NiHCFe are some of the PBAs that were studied. Their research revealed that CoHCFe-pp had the highest temperature coefficient of the three, at 0.89 mV K^{-1}. CuHCFe and NiHCFe have

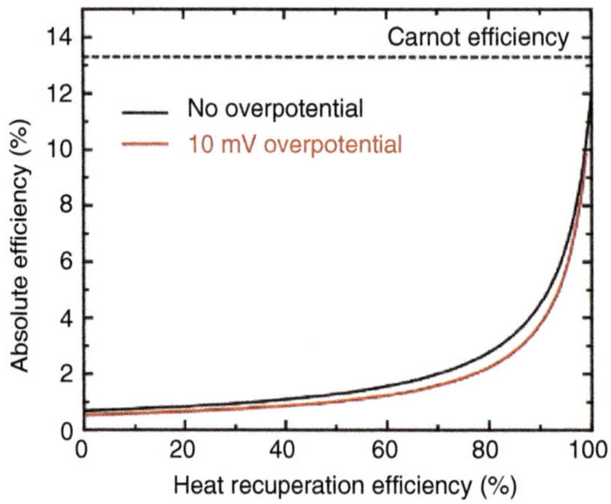

Figure 9.5 The absolute energy efficiency of the charging-free TREC vs. heat regeneration efficiency. Source: Lee et al. [9]/PANS.

temperature coefficients of 0.55 and 0.67 mV K^{-1}, respectively. According to reports, all of the PBAs under consideration had lattice shrinkage when cations intercalated. The CoHCFe-pp had a greater temperature coefficient than CuHCFe and NiHCFe because it had broader lattice parameters and a higher lattice parameter change ratio. Lattice alteration can have an impact on the temperature coefficient by influencing the change in phonon vibration entropy. CoHCFe-pp decided to use the TREC as a result. The TREC was charged at 50 °C and discharged at 10 °C. For a CoHCFe-pp-based TREC, the voltage against time (temperature) and specific capacity are depicted in Figure 9.6a,b, respectively [10]. The efficiency of the TREC system was found to be 2.65%, 3.65%, and 4.31% for the regeneration efficiencies of 0%, 50%, and 70%. These numbers correspond to corresponding ZTs of 1.6, 2.6, and 3.7. About 21% of Carnot efficiency was reached with a lesser efficiency for this temperature shift. After 50 cycles, the TREC was also steady.

A TREC based on CoHCF was studied by Jiang et al. Helical carbon nanotubes (HCNTs) are employed to create a composite of CoHCF/HCNTs that enhances the microtexture and performance of CoHCF. The temperature coefficient of CoHCF alone was reported to be 0.69 mV K^{-1}, but when HCNT was added, the temperature

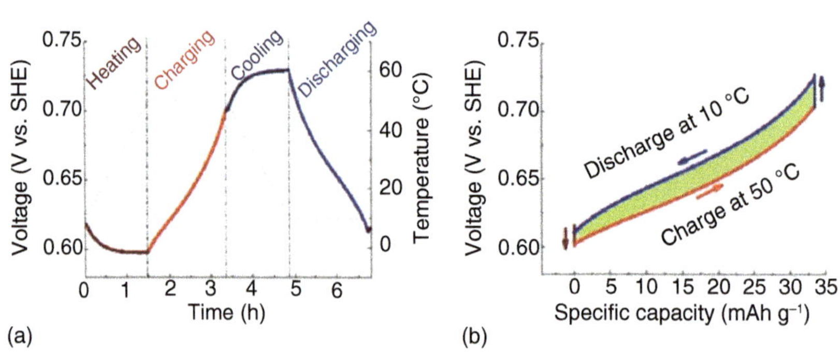

Figure 9.6 Diagram of (a) voltage curve via time and temperature, (b) voltage-specific capacity of the CoHCFe-pp based TREC system. Source: Gao et al. [10]/John Wiley & Sons.

Table 9.1 Comparison of the efficiency of CoHCF and CoHCF/HCNTs TREC systems in the temperature range of 15–45 °C for different regeneration efficiencies.

Regeneration efficiency	0	50%	70%
CoHCF TREC efficiency	0.45%	0.77%	1.08%
CoHCF/HCNTs TREC efficiency	1.18%	1.91%	2.53%

Source: Adapted from Jiang et al. [11].

coefficient of CoHCF rose and was found to be 0.89 mV K^{-1} for the CoHCF/HCNTs. Additionally, the addition of HCNTs caused the specific heat capacity of CoHCF to drop from 2.33 to 2.17 J g^{-1} K. Besides, compared to CoHCF, CoHCF/HCNTs showed a larger capacity and lower resistance. In order to determine the TREC efficiency and enhance its performance, experiments were carried out utilizing HCNTs. As a result, two CoHCF and CoHCF/HCNTs electrodes were used as working electrodes, while Ag/AgCl was used as the opposing electrode. Two temperatures of 15 and 45 °C were used to test TREC. In Table 9.1, the efficiency result is summarized. As can be observed, the CoHCF/HCNTs TREC system efficiency outperformed the CoHCF TREC system throughout a range of regeneration efficiencies. In this temperature range, the CoHCF/HCNTs TREC system's greatest efficiency was 27% of the Carnot efficiency. The results of the experiment further showed that the specific capacity of CoHCF/HCNTs did not decline and was steady after 10 cycles (about 35 000 seconds).

Reynard et al. studied the utilization of a vanadium redox flow battery (VRFB) in a TREC cycle in response to the studies performed on the TRECs [12]. These batteries use pumps to provide the electrolyte solution to the battery when it is being charged or discharged to produce or get power. Powerful devices, RFBs may resist tens of thousands of cycles and persist for decades. These batteries, however, have poor efficiency, low energy density, and low power density. Some of these limitations can be eliminated, and the battery performance can be enhanced, by merging the TREC with RFB. The reactions in a VRFB with the common electrolyte solution of these batteries are as follows:

$$(+)\text{side: } VO_2{}^+ + e^- + 2H^+ \rightarrow VO^{2+} + H_2O, E^0 = 1.00\text{ V} \tag{9.9}$$

$$(-)\text{side: } V^{3+} + e^- \rightarrow V^{2+}, E^0 = -0.26\text{ V} \tag{9.10}$$

In this work, a mixed-acid electrolyte was also employed in addition to the typical electrolyte of VRFBs, as seen in Figure 9.7. For the common electrolyte, the battery's positive and negative temperature coefficients were −0.151 and 1.013 mV K^{-1}, respectively. The entire cell temperature coefficient was thus measured to be −1.16 mV K^{-1}. It was also noted that the temperature coefficient for the cell using mixed-acid electrolyte was −0.80 mV K^{-1}. The low temperature was 20 °C, while the maximum temperature was thought to be 60 °C. Their research demonstrates that, in the usual electrolyte situation, the addition of the TREC cycle to the VRFB boosted the system energy density by 1.3 Wh l^{-1} in comparison to the VRFB without

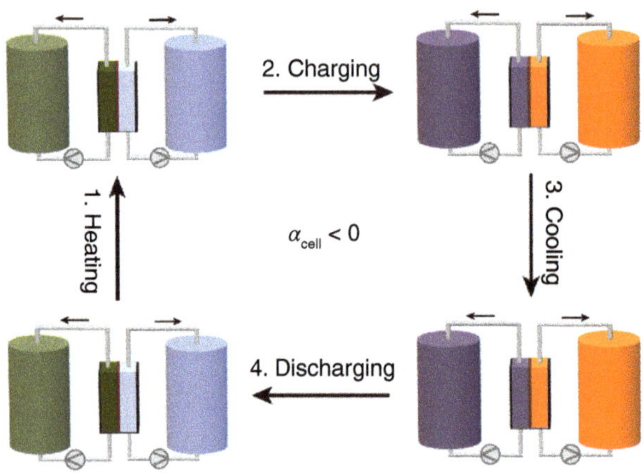

Figure 9.7 Schematic illustration of a vanadium redox flow battery combined with a TREC cycle (VRFB-TREC).

the TREC cycle. Additionally, energy efficiency has risen by 9%. These gains were 0.8 Wh l^{-1} and 5%, respectively, for the mixed-acid electrolyte situation. The TREC efficiency in this investigation for common and mixed-acid electrolyte between the two temperatures of 20 and 60 °C was 2.6% and 1.7%, respectively, which was 21.6% and 14.1% of Carnot efficiency, respectively. They also note that a solar farm connected to the VRFB-TREC can boost energy gathering by 10%. The VRFB-TREC may be charged with the heat generated by the farm during the day and recharged with electricity at night when it is cold, completing the TREC cycle.

Linford tested a lithium-ion-based TREC as part of his study on lithium-ion power-generating systems. Li–Al acted as the anode and LiV_2O_5 as the cathode in the TREC cell under investigation. LiV_2O_5 had a temperature coefficient of 1.85 mV K^{-1}, whereas Li–Al had a coefficient of 1.50 mV K^{-1}. This resulted in a positive cell temperature coefficient of 0.35 mV K^{-1}. Figure 9.8 displays the TREC cell's voltage capacity. Charged at 22 °C, the test cell was discharged at a higher temperature of 60 °C. In total, 1.89 mJ were collected from the cell throughout its half-hour of operation. Additionally assessed was a TREC made of lithium cobalt dioxide. $LiCoO_2$ (LCO) served as the anode and had a temperature coefficient of 1.70 mV K^{-1}, while Li served as the cathode. The cell temperature coefficient was thus 0.50 mV K^{-1}. It was discovered that LCO cells could transform heat energy into electrical work when the cell was discharged at normal temperature and charged at a higher temperature of 60 °C.

9.3 Combination of TRECs with Other Techniques

The usage of photovoltaic (PV) panels has recently attracted a lot of interest, and significant effort has been done to maximize and recover their lost heat [13, 14]. In an experimental investigation, Fathabadi [15] explored the usage of the continuous TREC system to raise a solar power generating system's output power efficiency.

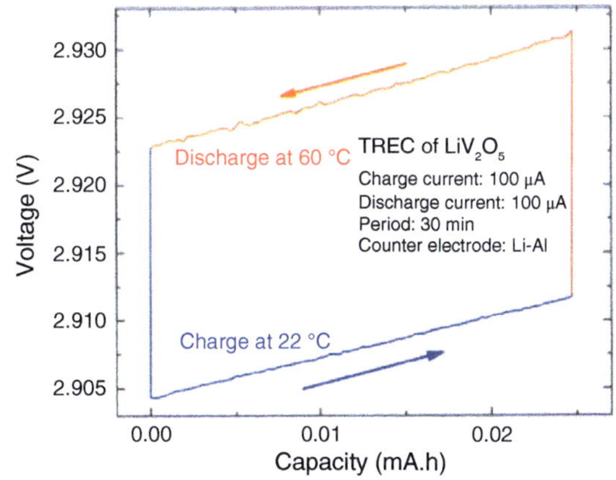

Figure 9.8 The voltage capacity of a LiV_2O_5-based TREC.

The substantial heat that solar modules waste is one of the main problems with PV power systems. Therefore, a continuous TREC system together with a 200 W PV module PV power generating system was combined for the first time to create a new hybrid system for harvesting this lost heat.

CuHCF in 6 M $NaNO_3$ is utilized as the positive electrode in the continuous TREC system suggested in this investigation, as in the study of Lee et al. [7], while Cu in 3 M $Cu(NO_3)_2$ is employed as the negative electrode. The continuous TREC system's temperature coefficient was 1.22 mV K^{-1}. The TREC system's hot cell is directly linked to the PV module's back, while its cold cell is attached to a heat sink that is hidden by the PV module. The continuous TREC system employed in this investigation weighed 3.1 kg. The price of the integrated system was similarly about 125 €. The cold source's temperature ranged from 296 to 314 K, whereas the hot source's temperature ranged from 301 to 365 K. According to experimental findings, the TREC system's power production ranges from 0.1 to 55 W, with an average of 27.3 W. This indicates that 1 kW of electricity is produced by the TREC system once every 37 hours. The system's measured efficiency ranged from 0.2% at its lowest value to 6% at its best. As a result, the TREC system's average efficiency was 3.1%, which translates to a 3.1% average for PV's recovered heat. A comparison of the PV system's daily power production with and without TREC was also made. When coupled with a TREC system, the PV system's power production grew by around 27.5% to reach 1.213 kWh day^{-1}, from the TREC-free PV system's average daily output of 0.952 kWh day^{-1}.

In a different research, Fathabadi [16] employed the same continuous TREC system for waste heat recovery in the vehicle's internal combustion engine and converted it to electricity. This system had previously been used for solar module heat recovery. The automobile battery subsequently housed this electrical energy. The TREC system used in this investigation weighed 3.8 kg and had the following measurements: $50 \times 30 \times 38$ cm^3. It is simple to utilize in an automobile engine because of its size and weight. The TREC system's hot cell and cold cell were both directly linked to the engine and radiator, respectively, of the vehicle. An image of this mechanism within the diesel engine may be found in Figure 9.9.

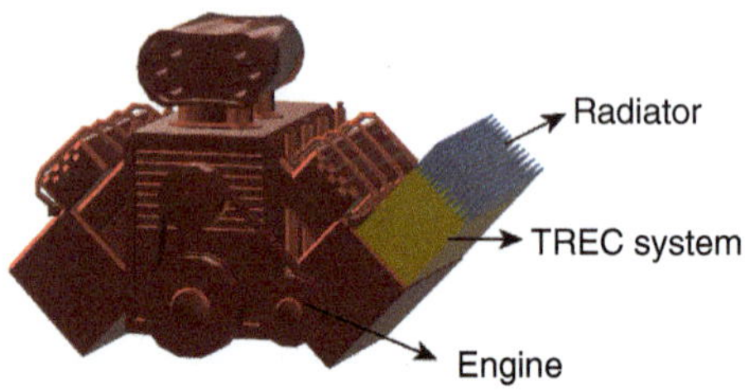

Figure 9.9 Schematic diagram of the continuous TREC system is used in the internal combustion engine of a vehicle.

When the engine was running, the hot cell ranged in temperature from 368 to 383 K, while the cold cell ranged from 318 to 333 K. According to experimental findings, the TREC system's power output, which was used to charge the vehicle's battery, ranged from 24.25 to 96.25 W, with an average power output of 60 W. This means that after 17 hours, the TREC system generated and stored more than 1 kW of electricity. The system's measured efficiency ranged from 4.1% at its lowest value to 7.2% at its best. Thus, the TREC system's average efficiency was 6%, which indicates that the average amount of engine heat it was able to collect was 6%. The whole system cost about 115 €. The initial purity of the positive and negative electrodes was roughly 98%, they added. The purity dropped to 60% due to deposition and an increase in the internal resistance of the cell over time, necessitating the replacement of the electrodes and electrolytes. According to reports, the electrodes and electrolytes had a lifespan of roughly four years on average.

In another study, Fathabadi [17] directly put the continuous TREC system on a building's roof. It may be viewed as an alternative to PV panels because its goal was to directly use solar energy and transform it into electricity. The TREC system requires that the hot cell, which is heated by solar energy, be directly attached to the hot section of a structure, such as an iron roof or window, and that the cold cell be situated in the building's shadowed area. In this experiment, the hot cell was positioned on a black iron roof of a building, while the cold cell was put in the cold region of the roof. Figure 9.10 depicts a general view of the TREC system installed on the building's top. The peak temperature of hot cells ranged from 300 to 367 K on a bright summer day, whereas the low temperature of cold cells ranged from 296 to 317 K.

According to test results, the power output and efficiency of a solar-powered TREC system ranged from 0.1 to 54.5 W and 0.2% to 6%, respectively. On a bright spring day with an ambient temperature ranging from 289 to 301 K, TREC's daily

Figure 9.10 TREC system is used on the building roof for using solar energy. Source: Fathabadi [17]/with permission from ELSEVIER.

Table 9.2 Comparison of TREC system and a photovoltaic module for power generation.

Parameters	Cost per kWh	Weight	Size	MPPT unit	Energy converted
TREC	317.16 €	3.1 kg	$50 \times 30 \times 30\,cm^3$	No need	Solar heat
PV module KC200GT	315.13 €	18.5 kg	$142.5 \times 99 \times 3.6\,cm^3$	Need	Solar irradiance

Source: Adapted from Fathabadi [17].

Table 9.3 Comparison of TREC system and thermoelectric generator for solar thermal harvesting.

Systems	Output current at 20 °C (A)	Output current at 50 °C (A)	Output power at 20 °C (W)	Output power at 50 °C (W)
TREC	247.5	877.5	6.2	54.5
TEG1-1263-4.3	0.11	0.24	0.04	0.3

Source: Adapted from Fathabadi [18].

electricity was also measured and contrasted against a PV module called the KC200GT. In Table 9.2, the findings of this comparison are displayed. The TREC system and PV module KC200GT produced roughly 0.268 and 0.952 kWh of power each day, respectively. While the PV module KC200GT costs 300 €, the TREC system only costs 85 €. Cost per kWh was therefore taken into account while comparing these two systems. The TREC system's and the KC200GT PV module's cost per kWh were roughly 317.16 € and 315.13 €, respectively. This comparison demonstrates unequivocally that the TREC-based device's economic parameter was marginally greater than the PV module's, despite the TREC's smaller size and lower weight.

Another research by Fathabadi looked toward replacing commercial thermoelectric generators (TEGs) with a continuous TREC system for low-temperature applications [18]. The fundamental drawback of a commercial TEG is its poor power production when the temperature differential between its surfaces is less than 50 °C, which makes it useless for most purposes, including solar energy harvesting. As a result, extra, expensive equipment like an evaporator, concentrator, and condenser are required to raise the temperature between its two surfaces. TREC was presented as a fresh approach to capturing solar energy in this study. The constructed TREC device had a $50 \times 30 \times 32\,cm^3$ dimension and a weight of 3.1 kg. The tests' findings demonstrated that the TREC system's output power could reach 54.5 W at temperature differences between 5 and 50 °C and that its power efficiency changed linearly with temperature at these temperatures from 0.2% to 6%. The thermoelectric module and the TREC device were compared in Table 9.3. The TREC produced 54.5 W of power at a temperature differential of 50 °C, compared to the thermoelectric module's (TEG1-1263-4.3) meager 0.3 W, as indicated in Table 9.3. The output power of the two systems was also evaluated at 20 °C, showing that the TREC system had a significantly greater power output than the thermoelectric module.

Table 9.4 Comparison of TREC system and TEG for converting waste heat of PV module to electricity.

Systems	Power production (at 13:00) (W)	Power reduction compensation (%)	Overall cost (€)	Size (cm^3)	Weight (kg)
TREC	45	85	210	50 × 30 × 32	3.1
TEG-DW-WC-100W	12	23	950	60 × 14.6 × 5.7	18.5

Source: Adapted from Fathabadi [19].

In another experimental study by Fathabadi [19], a continuous TREC system and the water-cooled DW-WC-100 W TEG module were employed as two separate systems to make up for the decreased power production of a PV module (type: KC200GT) when its surface temperature rose. Both systems were linked to the PV module back in order to convert the waste heat from the PV module into energy. The linked system of a continuous TREC system for a day was the subject of experimental studies. The output power of the entire system, the PV module, and the continuous TREC system were 192, 147, and 45 W, respectively, at midday (during the time of maximum solar irradiation), while the ambient temperature, the surface temperature of the PV module, and the heat sink temperature were, respectively, 298, 349, and 302 K. Additionally, as reported, the output power of the PV module reduced from 200 to 147 W (about 53 W) as the temperature climbed from 298 to 349 K. As a result, the continuous TREC system has compensated for a loss in output power of the PV module of around 85% (45 W/53 W). The whole system and PV module output power were 159 and 147 W, respectively, at midday when the TEG module was employed. As a result, the TEG module generated around 12 W, which is obviously less than the continuous TREC system. The TEG module DW-WC-100 W has only adjusted for a 23% reduction in PV module output power. Table 9.4 contrasted the continuous TREC system with the TEG module DW-WC-100 W. As can be seen, the TREC system has a higher output power and is lighter and less expensive than a TEG module.

9.4 Challenges and Outlooks

For power generation and refrigeration applications, the benefits of TREC and thermally regenerative electrochemical refrigeration (TRER) (the reverse cycle of TREC for cooling applications) systems include their quiet operation, low weight and volume, cheap cost, lack of moving elements like a compressor, clean technology, and flexibility to transition from chilling to heating with the same equipment. The ability of the electrochemical refrigerator to be connected to any voltage source, such as solar panels, and the applicability anywhere waste heat may be converted into electricity are further advantages. To overcome these obstacles, TREC and TRER electrochemical systems must be used in future projects. Also included in this section are some recommendations for more research and potential applications:

The choice of appropriate redox pair materials is the first item to take into account when using the TREC and TRER systems. Selecting a redox pair with a high Figure of merit, $y = \alpha c_q/c_p$, is essential for creating a high-efficient TREC. When choosing a material, the three factors of temperature coefficient, heat capacity, and charge capacity must be taken into consideration together with quick kinetics. Solid electrodes having an open nanoscale structure, such PBAs nanoparticles, can be employed for rapid kinetics. A high-temperature coefficient is necessary for a high-performance TREC, but just having high-temperature coefficient materials does not ensure these systems will operate at their peak. An all-inclusive list of the temperature coefficients of reactions can be found in a study done by de Bethune et al. Therefore, further theoretical research is needed to better understand the temperature coefficient and the Figure of merit in electrochemical systems. Ions in the electrolyte, phonon entropy in the solid phase, configurational entropy, electron entropy, and other entropy changes are only a few of the entropy changes that contribute to the entropy shift in a chemical process. Additionally, according to the research of Gao et al., the temperature coefficient of PBAs is closely related to the alteration of the lattice parameter, which is mostly linked to the various phonon vibration entropy variations. Therefore, a thorough knowledge of the link between entropy change and the characteristics of the reaction materials is crucial to the development of a high-efficiency TREC. It should be noted that the criteria that make up the TRECs' Figure of merit are not connected to one another. Another benefit of TREC systems over thermoelectric systems is that they may be adjusted independently to optimum efficiency. To create a new TREC with high efficiency, novel redox pairs with high-temperature coefficient, charge capacity, and low heat capacity must be found.

A recently created electrochemical cell is made up of a CuHCF cathode and a manganese hexacyanomanganate open-framework anode. The electrochemical cell that was presented performed in a safe, quick, affordable, and long-cycle life manner. After 1000 deep discharge cycles, there is no capacity loss. In a different study, Pasta et al. suggested using manganese and CoHCF in an aqueous sodium-ion electrolyte to increase the specific capacity of PBA materials. These materials have features like rapid kinetics, high specific capacity, high energy efficiencies, and long cycle lives in comparison to previously reported PBAs. As a result, these redox pairings, which have never been utilized to create TRECs, may be viable choices for inclusion in TREC systems.

The number of cycles that the TREC and TRER systems can function is the other factor to take into account. After a number of cycles, reaction materials may collect, the utilized membrane may lose its effectiveness, and this will increase the cell's resistance. Ion diffusion across the membrane and direct redox reactions on the membrane's sides are both possible effects of membrane efficiency loss. As a result, choosing advanced membranes with zero or extremely low permeability should be taken into account. The membrane's surface coating may improve its effectiveness. The reactant materials also run out after a number of working cycles; thus it is important to think about how to inject or recharge the reactant materials needed by the continuous system. Therefore, it is important to consider how many cycles or how long the system can work at peak efficiency, and it is also wise to use materials that can support more cycles. For TRECs to be commercially viable, the cycle life would need to be more than 10 000.

The output power and efficiency of the TREC system are greatly impacted by regeneration efficiency, as the literature study shows. It is not clear how to accomplish extremely effective heat regeneration due to the tiny temperature difference between the hot and cold cells and charging and discharging at a constant temperature. In addition, hot and cold sources are in direct contact with the hot and cold cells, which makes them ideal heat exchangers in theory. However, in practice, direct contact between the hot and cold sources and the hot and cold cells may not be possible and requires materials with high thermal conductivity. Since the heat regeneration mechanism between cold and hot cells is still a mystery, creating high-efficiency and affordable heat exchangers for hot and cold cells is also a challenge. No practical work has been done to support this claim, despite the fact that several theoretical papers suggest that TREC may be used to effectively harvest the waste heat from many types of fuel cells. The practical use of TREC with fuel cells and the analysis of its economic aspects are thus some recommendations for further research.

The combination of a hydrogen/oxygen fuel cell and a water electrolyzer, as studied by Newell, demonstrated a promising cooling effect; however, the system may not be suitable due to the high activation potential of the oxygen electrode; as a result, identifying and substituting the oxygen reaction may be the answer for a hydrogen/oxygen fuel cell and water electrolyzer-based TREC in practical applications. In this area, computational science can be useful. The TREC system's performance may be assessed in any cycle and system where waste heat is present since it can utilize waste heat to produce electricity and function as a secondary power generator. As seen, TREC enhanced the energy efficiency of these systems by integrating with other renewable energy systems and cycles. Many attempts have been made to increase the thermal efficiency of various systems and cycles, particularly the gas-turbine cycle and the turbo-compressor gas engine. While this heat may be utilized in a TREC system, heat recirculation is thought to be a technique to increase the power and thermal efficiency of these cycles. Therefore, one of the recommendations for further study is to compare the TREC to comparable systems in order to boost their thermal efficiency.

The TREC and TRER may be used in cogeneration and polygeneration applications, respectively. By using technology for the recovery of waste heat of energy, TREC can reach combined heat and power applications. The study of TREC-CHP-based systems is advised for future efforts. It is advised to research and examine the suggested TREC-TRER-PV polygeneration system for power generation, heating, and cooling. For internal combustion applications, Izadiamoli and Sayyaadi employed an Otto-Stirling engine/cooler hybrid system to recover waste heat thermal energy from exhaust gasses. The TREC technology may potentially be used in automotive applications to recover heat. In light of the automobile industry's shift to cost-effective goods, future research might compare these two cycles.

Although the TRER cycle has been well researched in theory, more experimental research on continuous TRER cooling systems will be necessary. The temperature control of electric car battery systems is one of the uses for continuous TRER. TRER may be employed as a potential solution for cooling the batteries of electric cars since they have recently needed a battery cooling system with attributes like silent performance, low volume, and optimal efficiency.

In today's energy discussion, the economic component is a crucial consideration when deciding on systems for both power generating and cooling. Future research should focus on two crucial areas: finding novel low-cost materials and creating low-cost ion-selective membranes. The supply of reactive materials, electrolyte materials, membrane costs, cell electrodes, and materials required to build TREC systems cells for practical applications are only a few of the TREC and TRER's economic considerations that need to be addressed. Because this cycle is a new system, and as evidenced by the literature review, relatively few research has looked into its economic elements. As a result, the gap in economic research on these systems may be shown theoretically or empirically. Further study on the economic implications of these systems, based on the life cycle assessment, is recommended for future studies in order to ensure their broad usage.

As can be concluded, multidisciplinary research involving material science, chemistry, and chemical engineering is required to develop chemical features of TREC, such as novel materials, redox pairs, electrodes, membranes, life cycle, and temperature stability of the materials. Furthermore, mechanical engineering investigations need the development of innovative system designs. Computational science studies can aid in both material and system design for future TREC system research.

References

1 Zebarjadi, M., Esfarjani, K., Dresselhaus, M.S. et al. (2012). *Energy Environ. Sci.* 5: 5147.

2 Snyder, G.J. and Toberer, E.S. (2008). *Nat. Mater.* 7: 105.

3 Poudel, B., Hao, Q., Ma, Y. et al. (2008). *Science* 320: 634.

4 Salazar, P.F., Kumar, S., and Cola, B.A. (2012). *J. Electrochem. Soc.* 159: B483.

5 Gunawan, A., Lin, C.-H., Buttry, D.A. et al. (2013). *Nanoscale Microscale Thermophys. Eng.* 17: 304.

6 Abdollahipour, A. and Sayyaadi, H. (2021). *Appl. Therm. Eng.* 187: 116576.

7 Lee, S.W., Yang, Y., Lee, H.-W. et al. (2014). *Nat. Commun.* 5: 3942.

8 Wessells, C.D., Huggins, R.A., and Cui, Y. (2011). *Nat. Commun.* 2: 550.

9 Yang, Y., Lee, S.W., Ghasemi, H. et al. (2014). *Proc. Natl. Acad. Sci. USA.* 111: 17011.

10 Gao, C., Yin, Y., Zheng, L. et al. (2018). *Adv. Funct. Mater.* 28: 1803129.

11 Jiang, J., Tian, H., He, X. et al. (2019). *J. Mater. Chem. A.* 7: 23862.

12 Reynard, D., Dennison, C.R., Battistel, A., and Girault, H.H. (2018). *J. Power Sources.* 390: 30.

13 Sohani, A. and Sayyaadi, H. (2020). *Renew. Energ.* 156: 395.
14 Sohani, A. and Sayyaadi, H. (2020). *Energ Convers. Manag.* 224: 113291.
15 Fathabadi, H. (2019). *IEEE Trans. Power Electron.* 34: 8626.
16 Fathabadi, H. (2018). *IEEE Trans. Veh. Technol.* 67: 9241.
17 Fathabadi, H. (2019). *Renew. Energ.* 133: 118.
18 Fathabadi, H. (2019). *Energy.* 174: 932.
19 Fathabadi, H. (2019). *Renew. Energ.* 142: 543.

10

Integration of Energy Conversion and Storage Devices

10.1 Introduction

The high value-added utilization of low-grade heat is still plagued because of the lack of promising energy conversion and storage devices under a small temperature gradient between the hot and cold sides [1]. Different from the low Seebeck coefficient of traditional electronic thermoelectric (e-TE) devices and relatively complex integration of Organic Rankine Cycle (ORC), liquid thermocells (LTCs) hold great promise for low-grade heat conversion due to their unique merits such as cost-effectiveness, good flexibility, high Seebeck coeffici ents, and power density [2]. Inspired by heat-to-current mechanisms, two kinds of LTCs based on the electric double-layer capacitive processes and pseudocapacitive reactions have attracted much attention in realizing the high value-added conversion of low-grade heat [3]. Notably, the output voltage and the conversion efficiency of the reported LTCs are still low, which could mainly be determined by their near-surface energy conversion procedures. Thus, it is required to develop and construct a new-type ionic TE (i-TE) device with fast heat/electricity response to keep pace with the demand of the Internet of Things.

Alternatively, hybrid ion capacitors (HICs) are expected to achieve both high energy density and power density due to the reasonable integration of batteries and capacitors. In particular, nowadays, zinc-based devices have been regarded as one of the potential candidates for scalable energy storage, which can be attributed to the low redox potential (−0.76 V vs. the standard hydrogen electrode), high theoretical capacity (820 mA h g^{-1}/5855 mA h cm^{-3}), and satisfying water media compatibility of the Zn metal [4]. When combining the concept of Zn-related devices and heat-to-current conversion, a new-type i-TE proposed by our research group, namely, zinc ion thermal charging cell or thermoelectrochemical cell (ZTEC), can convert low-grade heat into electricity through the synchronous thermodiffusion and thermo-extraction/insertion of electrolyte ions in the battery-type cathode and stripping/plating behavior of the Zn anode (Figure 10.1a) [3, 5]. It is well known that lithium-ion battery (LIB) also is one of the important nonaqueous candidates for energy storage applications owing to its inherent merits such as high energy, high power, and ultralong cycle life. With the development of electrode materials for Li$^+$ storage, various charge storage mechanisms including

Low-Grade Heat Harvesting: Materials, Devices, and Technologies, First Edition. Xiaogang Zhang.

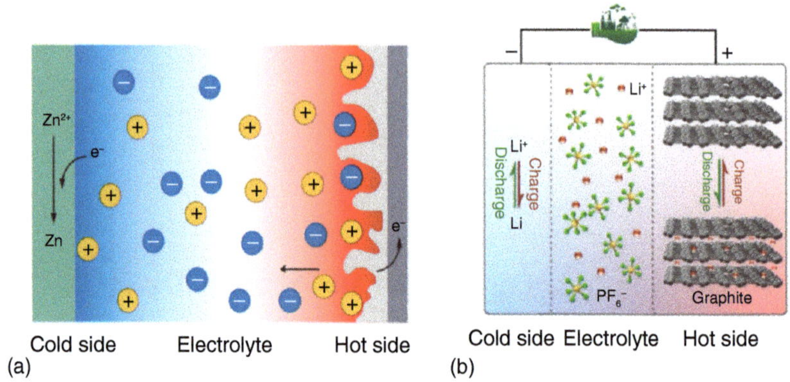

Figure 10.1 Schematical illustration of (a) ZTSC and (b) LITC. Source: Adopted from Li et al. [3].

intercalation/deintercalation and phase transition have been proposed. Inspired by the reversible energy storage process and thermodynamic response of ions, the lithium-ion thermal charging cell (LITC) can be reasonably developed by applying the temperature gradient (Figure 10.1b) [6].

10.2 Mechanisms

HICs as a new and most attractive low-quality heat collection system can effectively combine the properties of electrodes and electrolytes. Most importantly, the HICs built by embedded materials (such as $LiMn_2O_4$), electric double layer capacitive materials (such as PC), and functional electrolytes can realize energy conversion and storage at the same time. Generally, the cations embedded in the positive electrode can be removed under the temperature gradient and transported to the negative electrode under the thermal field drive. In the process of ion transport, the electrons generated by the oxidation–reduction reaction during the removal of cations can move from the hot end to the cold end through the external circuit. After the charge is accumulated on the negative surface, it will absorb a large number of cations. This leads to an increase in the difference in ion concentration between the cold and hot end electrodes, further increasing the thermal voltage of TEC. Therefore, the thermal deblocking process still belongs to the category of TE chemical conversion caused by pseudo-capacitance behavior. Similar to the development strategy of the above high-performance devices, porous electrodes and embedded electrodes should have rich electroactive sites, good conductivity, and high capacity, providing sufficient potential for ion storage and charge transfer.

In recent years, due to the unique characteristics of metal anodes, metal anodes with electrochemical deposition/stripping properties have become the most promising electrode materials for building batteries or capacitors for energy storage. An integrated device for energy conversion and storage can be constructed when

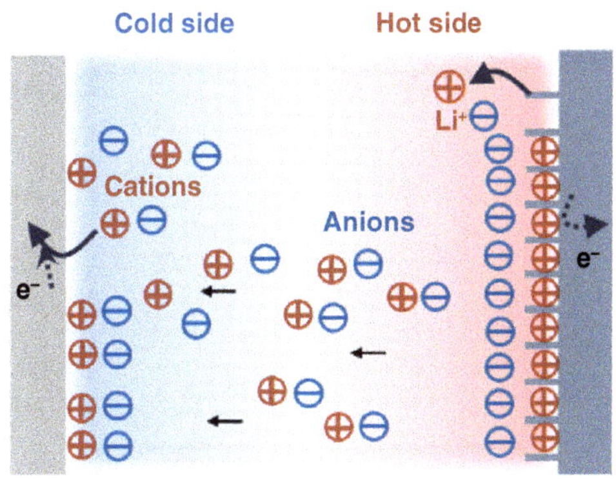

Figure 10.2 Working principle of TEC based on the embedded positive electrode and metal negative electrode.

the embedded cathode material is matched with the cheap metal anode (such as zinc). As shown in Figure 10.2, the cations embedded in the positive electrode through electrochemical discharge can be gradually removed with the continuous input of heat at the hot end. After these cations are transported to the negative side, they are deposited on the surface of the metal electrode through electrochemical process. In the whole TE chemistry process, the generated electrons need to be transferred from the hot end to the cold end through the external circuit to realize the conversion of low-quality heat energy to electric energy and the storage of electric energy. In addition, TECs based on metal anode can achieve high TE conversion efficiency and energy storage density. It is worth mentioning that the metal anode mentioned here can also be replaced by other promising candidate materials, such as lithium, sodium, potassium, and some alloying anodes. However, when using metal electrodes that are incompatible with aqueous electrolyte, the nonaqueous electrolyte engineering and its corresponding device integration technology should be fully considered to avoid possible safety problems.

The low-quality thermal-electrical conversion device based on thermal stripping/embedding effect is a new type of energy conversion and storage system. The design principle of TECs is similar to that of rechargeable secondary batteries. Therefore, with the rapid development of advanced materials and battery technology, the research on TECs has also ushered in explosive growth.

As shown in Figure 10.3, Kobayashi et al. inspired by the structure of lithium/sodium ion batteries, proposed a new type of thermal battery (also known as battery type thermal battery) consisting of two bonded electrodes, and studied its corresponding TE chemical properties [7]. In this work, Kobayashi et al. employed $Na_{0.99}CoO_2$, $Na_{0.52}MnO_2Na_{0.5}1Mn_{0.5}Fe_{0.5}O_2$ and $LiCoO_2$ as working electrodes. In addition, the TE pressure of $Na_{0.99}CoO_2$ at a specific temperature difference has a strong time dependence. Therefore, the Seebeck effect of the battery thermal battery based on this electrode material may largely depend on the insertion and removal of Na^+.

In addition, the research team of Xuan has built a thermally charged hybrid supercapacitor (TCHS) based on an LMO/G/CNT cathode and PC anode, which converts low heat into electric energy [8]. Through the coupling of the electric double-layer

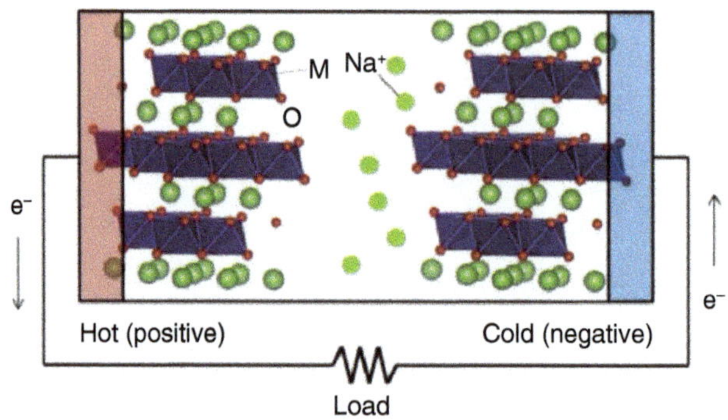

Figure 10.3 Schematic diagram of a battery-type thermal battery. Source: Kobayashi et al. [7]/AIP Publishing.

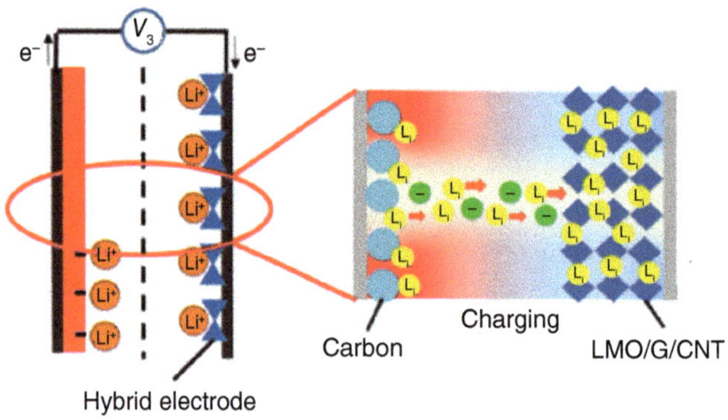

Figure 10.4 Schematic diagram of thermal charging hybrid supercapacitor. Source: Meng et al. [8]/American Chemical Society.

behavior and the intercalation effect, the high ion Seebeck coefficient of 12.1 mV K^{-1} and the output power density of 20.4 W m^{-2} can be achieved at a temperature difference of 50 K, which further proves the application possibility of thermal charging hybrid supercapacitors in low-quality heat recovery (Figure 10.4).

Compared with monovalent ion battery system such as $Li^+/Na^+/K^+/NH_4^+$, polyvalent ion ($Zn^{2+}/Mg^{2+}/Ca^{2+}$, etc.) battery has higher energy storage capacity. Metallic zinc has a high specific capacity (820 mAh g^{-1}/5855 mAh cm^{-3}), low standard electrode potential (−0.76 V), low cost (≈ \$65 kWh^{-1}), and high safety. Zinc ion batteries (ZIBs) are one of the most potential energy storage devices for large-scale application. Li et al. proposed a zinc ion TE chemical system for low-quality heat conversion and energy storage [5]. The TE chemical system can be divided into zinc ion TE chemical supercapacitors (ZTES) with PC and other adsorption/desorption materials as working electrodes (Figure 10.5a) and zinc ion TE chemical (ZTEC) batteries with cell-type materials such as VO as cathode (Figure 10.5b). ZTES based on ion thermal diffusion effect has relatively higher kinetic behavior than ZTEC based on thermal detachment/embedding effect, and ZTEC has more

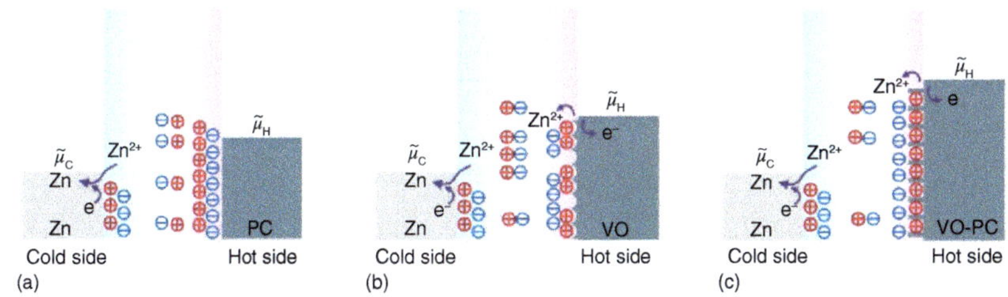

Figure 10.5 Working principle and application of zinc-based thermoelectric system: (a) thermal diffusion effect based on PC, (b) heat based on VO Detachment/embedding effect, (c) synergistic effect based on VO-PC.

compact charge storage capacity than ZTES. Therefore, by reasonably combining the thermal diffusion and thermal desorption/embedding behavior of ions in the electrode material, the constructed TE chemical device is expected to achieve high output voltage and power density at the same time. The possible mechanism is shown in Figure 10.5c. It is worth mentioning that zinc anode with a negative Seebeck coefficient ($-0.52\,\mathrm{mV\,K^{-1}}$) can further improve the TE chemical performance of zinc-based TE chemical system.

HICs are in the early stage of development, and their performance evaluation indicators are still established on the basis of existing knowledge. Therefore, it is very important to understand and develop a method to correctly evaluate the performance of HICs, which involves the measurement of heat collection output (Sorect coefficient, coefficient of merit, and charging efficiency) and electrochemical energy storage parameters (working voltage, specific capacitance, energy density, and power density). Compared with TEs equipment, the energy collection phenomenon of HICs has unique properties. It is similar to LTCs, and uses ions as carrier fluid, rather than electrons and holes as charge carriers.

TECs can be charged using intermittent heat sources. First of all, the heater is turned on to establish a ΔT. After a certain stabilization time, a thermal voltage value is generated, which is given by the following formula:

$$V_{\text{thermo}} = \alpha_i \Delta T \tag{10.1}$$

where α_i is the Seebeck coefficient of electrolyte (unit: $\mathrm{V\,K^{-1}}$) and ΔT is the temperature difference between the two electrodes (unit: K).

The thermal output power of HICs can be improved by adjusting the interaction between the ions and solvents that make up the electrolyte, because these interactions can change the difference between the heat-induced mobility of cations and anions.

The charge (Q_{ch}) stored on TECs is calculated by the following formula:

$$Q_{\text{ch}} = CV_{\text{thermo}} = C\alpha_i \Delta T \tag{10.2}$$

where V_{thermo} is the thermal voltage (unit: V), C is equivalent to the capacity of the equipment (unit: F), α_i is the Seebeck-like coefficient of electrolyte (unit: $\mathrm{V\,K^{-1}}$), and ΔT is the temperature gradient in K.

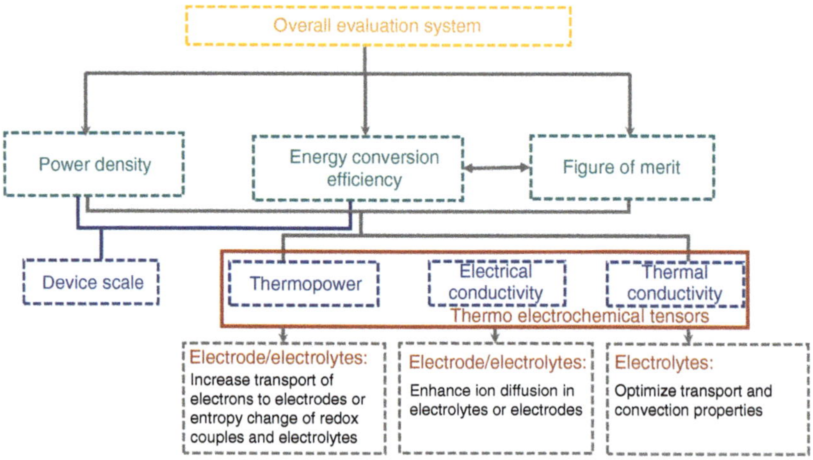

Figure 10.6 Illustration of components in the thermal charging cell that affect energy conversion efficiency, Figure-of-merit, and power density.

In addition, the electric energy stored in the equipment can be obtained in the following ways:

$$E_{ch} = \frac{1}{2}\frac{Q_{ch}^2}{C} = \frac{1}{2}CV_{thermo}^2 \tag{10.3}$$

This clearly shows that the stored electric energy increases squarely with the increase of the Seebeck-like coefficient of the electrolyte. If the current loss occurs, the equation for determining the stored electric energy needs to be restated as a function of the Seebeck coefficient of the electrolyte:

$$E_{ch} == \frac{1}{2}CV_{thermo}^2 = \frac{1}{2}C(\Delta T\alpha_i - V_s - V_{load})^2 \tag{10.4}$$

where ΔT is the temperature gradient in K, α_i is the Seebeck-like coefficient of the electrolyte (unit: $V\,K^{-1}$), C is the specific capacitance of the equipment ($F\,kg^{-1}$, $F\,m^{-2}$ or $F\,m^{-3}$ for a single device), V_s is the voltage drop on the internal series resistance (unit: V), and V_{load} is the voltage on the load resistance (unit: V).

Maximum charging efficiency of single HICs ($\eta_{\Delta Tch}$, unit:%) can be determined by (Figure 10.6):

$$\eta_{\Delta T_{ch}} = \frac{\Delta T}{2T_H + \frac{10k}{\sigma_i \alpha_i^2} - \frac{1}{2}\Delta T} \times 100 = \frac{\Delta T}{T_H} \frac{ZT_i}{2ZT_i + \frac{10T}{T_H} - \frac{1}{2}ZT_i\frac{\Delta T}{T_H}} \times 100 \tag{10.5}$$

10.2.1 Thermally Regenerative Electrochemical Cycles (TRECs)

Similar to thermogalvanic cells (TGCs) and thermally charged capacitors (TCCs), thermal regenerative electrochemical cycle (TREC) also has the ability to convert heat into electricity, but there are significant differences in structure and conversion mechanism compared to traditional TEC [9–11]. First, as shown in Figure 10.7a, a pair of asymmetric electrodes with positive and negative Si values are required in TREC to react with the electrolyte for energy exchange, which is different from the

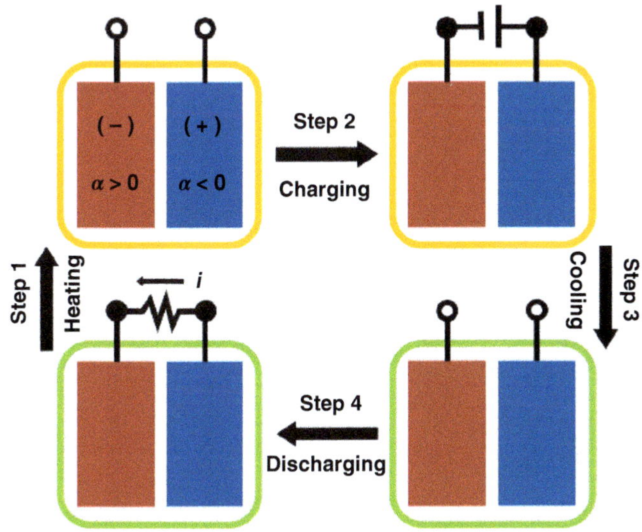

Figure 10.7 The mechanism illustration of the TREC, which undergoes a thermal cycling during operation: step 1, heating up the cell; step 2, charging at high temperature; step 3, cooling down the cell; and step 4, discharging at low temperature.

electrode only serving as a fluid collector in TGC [12]. The second difference between TREC and traditional TEC is that the energy conversion process of TREC usually consists of four thermal cycles: heating process, charging process, cooling process, and discharge process [12]. This means that TREC charges at high temperature and then discharges at low temperature. Alternatively, the charging process can be carried out at low temperature, followed by the discharge process at high temperature. These two different working mechanisms depend on the total thermopower (Scell) of TREC. When Scell is greater than zero, TREC charges at high temperatures, and when Scell is less than zero, TREC charges at low temperatures [10].

The network of TREC is energy output during the thermal cycling, due to the fact that the charging voltage is always lower than the discharge voltage during the thermal cycling process of TREC [10, 12] Figure 10.8 shows the temperature and entropy changes of TREC during thermal cycling. In the past decades, research studies about TREC have mainly focused on high-temperature conversion. Reaching the Carnot limit results in a thermal conversion efficiency of approximately 40–50% [12].

Using TREC for the conversion of low-grade heat conversion has been a new development direction in recent years. As shown in Figure 10.9, Lee's team developed a TREC with a thermal power of $-1.2\,\mathrm{mV\,K^{-1}}$ by using solid copper hexacyanoferrate (CuHCF) and Cu/Cu^{2+} as positive and negative electrodes in aqueous electrolytes [12]. In another recent work, Wang's team successfully prepared a TREC with a thermal power of $-5\,\mathrm{mV\,K^{-1}}$ using $Fe^{2+/3+}$ aqueous electrolyte and GO/Pt and PANI as positive and negative electrodes [13]. Figure 10.10a shows a new zinc ion thermal charging cell, to generate electricity from low-grade heat via the thermo-extraction/insertion and thermodiffusion processes of insertion-type cathode (VO_2PC) and stripping/plating behavior of Zn anode [5].

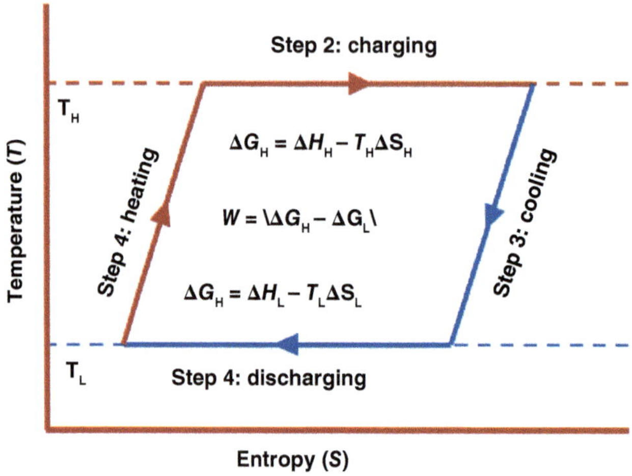

Figure 10.8 Temperature–entropy (T–S) diagram of the TREC during the thermal cycling.

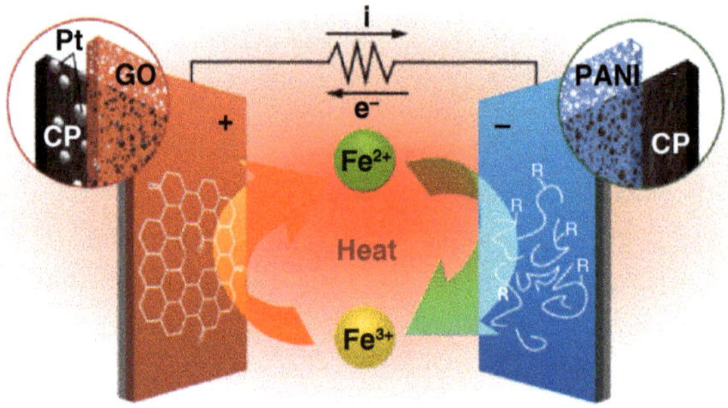

Figure 10.9 The GO/PtNPs//Fe^{2+}, Fe^{3+}//PANI/CP TREC proposed by Wang et al. schematic diagram, heat to electricity conversion mechanism. Source: Wang et al. [13]/Springer Nature/CC by 4.0.

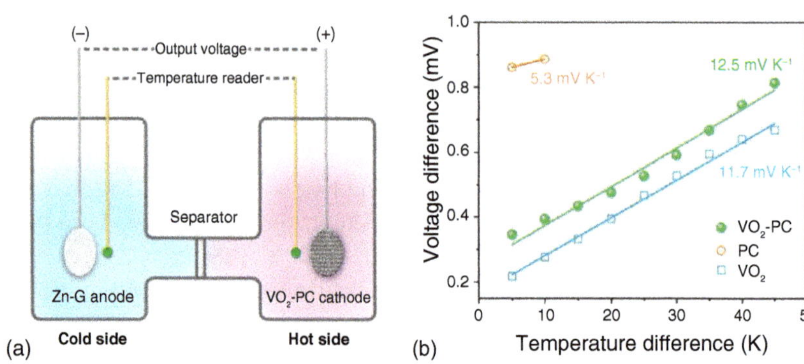

Figure 10.10 Zinc ion thermal charging cell assembled by Li et al.: schematic diagram (a) and thermopower (b) of this assembled cell. Source: Li et al. [5]/Springer Nature/CC by 4.0

10.3 Engineering Designs for Thermoelectrochemical Cells

In addition to improving the efficiency of single cells by improving TE energy or electrical conductivity, another solution worth further research is to optimize the structural design of TECs. Although TEC based on water-based electrolytes has shown great potential in TE conversion, water-based electrolytes still face many challenges. For example, liquid leakage and narrow temperature range. In order to solve these problems, researchers have developed quasi-solid-state electrolytes TEC system. Quasi-solid-state electrolytes can not only significantly improve the TE conversion efficiency of TEC but also give TEC some new characteristics and functions. For example, some hydrogel electrolytes and organic quasi-solid electrolytes can regulate the interactions between the ions and their surroundings. Due to these advantages, quasi-solid-electrolytes TEC systems have higher application potential.

10.3.1 p/n-Type Conversion

The p/n type conversion is one of the most critical parts in TECs system design, in other words, converting p-type TECs to n-type, or converting n-type TECs to p-type. Although some TECs systems reported recently have thermopower (Si) as high as dozens of mV/k, it is still challenging for single TECs with small temperature differences to power electronic devices at a working voltage at the level of volts. Serializing a large number of individual TECs is an effective solution to enable TECs systems to stably output voltage at the volt level, which requires the p-type TECs and n-type TECs mentioned earlier. From the current research status, the majority of studies on TECs focus on p-type, while there is relatively little research on n-type. In TCCs, the definition of p-type and n-type are related to the migration rate of anions and cations in the electrolyte. If the anion is faster than the cation, it is p-type, and if the cation is faster than the anion, it is n-type. Therefore, regulating the interaction between anions/cations and the surrounding environment can change the migration rate of ions, thereby achieving the conversion of p-type and n-type TCCs [14].

When anion thermal diffusion dominates in 1-ethyl-3-methylimidazolium (EMIM): bis(trifluoro-methylsulfonyl)imide/poly(vinylidene fluoride-co-hexafluoropropylene) (TFSI/PVDF-HFP) polymer gel, it exhibits an Si value of $-4\,mV\,K^{-1}$ (n-type) (Figure 10.11a) [15]. However, when polyethylene glycol (PEG) is doped in the polymer gel, Si value becomes $14\,mV\,K^{-1}$ (p-type) (Figure 10.11c). This doping can realize the thermal diffusion domination of cations through the interaction between cations and PEG (Figure 10.11b). Chi's team used the same polymer as the gel and ionic liquid Na: TFSI as the energy carrier, and added tris(pentafluorophenyl)boron (TPFPB) to the gel. The Si value changed from 20 to $-6\,mV\,K^{-1}$, that is, from p-type to n-type, effectively capturing the migration of Na^+ ions [16]. Liu's team reported in another work that a giant and bidirectionally tunable Si value can be generated by ion doping in non-aqueous ionogels [14]. In this work, thanks to the anion-dominated thermal diffusion, the reported Si value of EMIM: TFSI/PVDF-HFP ionogel is $-4\,mV\,K^{-1}$, (n-type). In addition, adding ionic

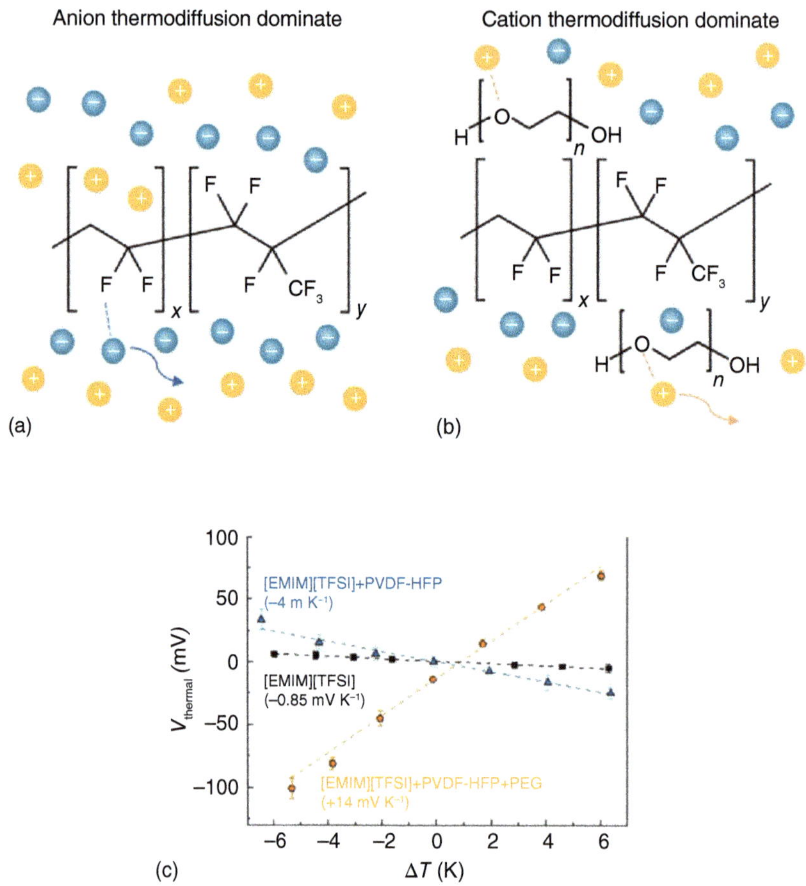

Figure 10.11 (a) Schematic illustration of the interaction between ions and the matrix before the addition of PEG. (b) Schematic illustration of the interaction between ions and the matrix after the addition of PEG. (c) The thermovoltage dependence on temperature difference with and without PEG in ionogels. Source: Zhao et al. [15]/Springer Nature/CC by 4.0.

liquid Li:BF_4 can adjust ion mobility and further increase Si value to $-15\,\mathrm{mV\,K^{-1}}$, (n-type). The addition of Li:BF_4 makes ion cluster $[\mathrm{Li(BF_4)}_m\mathrm{(TFSI)}_n]^{1-m-n}$ form in the ionogel, which will slow down the migration rate of cations and promote the migration of anions (Figure 10.12a). In addition, using ionic liquid EMIM:Cl instead of ionic liquid Li:BF_4, due to the opposite ion cluster effect, the migration rate of anions will be lower than that of cations (Figure 10.12b). Therefore, it will change to p-type with a Si value of $17\,\mathrm{mV\,K^{-1}}$.

In addition, whether it is p-type or n-type, their Si value can be adjusted by adjusting humidity, with a specific range of −23 to $32\,\mathrm{mV\,K^{-1}}$. Compared to the research studies in TCCs, there is only a very limited amount of research on the conversion process between p-type and n-type in TGCs. Duan's team reported the only case of p/n type conversion in TGCs so far. In this work, the author team introduced

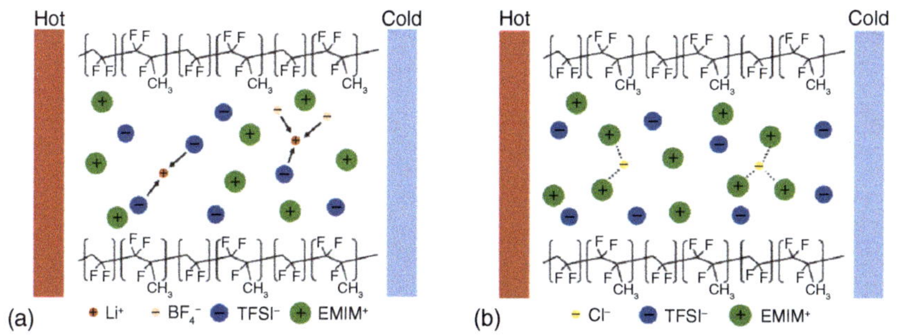

Figure 10.12 (a) Schematic to illustrate the interaction between Li^+ cations and anions during the thermodiffusion of ions in the polymer channel. (b) Schematic to illustrate the interaction between Cl^- anions and cations during the thermodiffusion of ions in the polymer channel.

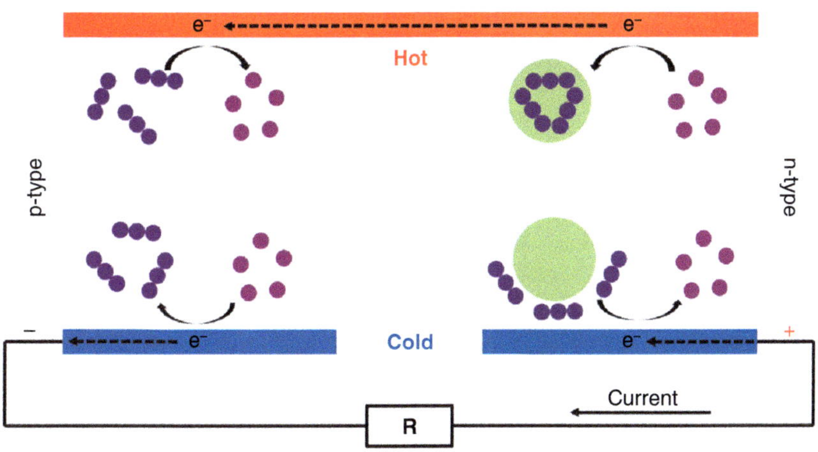

Figure 10.13 Host-guest design: schematic of the p/n conversion mechanism.

thermo-sensitive nanogels poly(N-isopropylacrylamide) (PNIPAM) into the I/I_3 TGCs system (Figure 10.13) [17]. As the PNIPAM nanogels have hydrophobic effect, they can capture I_3 on the hot side and release I_3 on the cold side through the swelling phenomenon. Therefore, the redox couple of I/I_3 can realize the transition from n-type ($0.71\,mV\,K^{-1}$) to p-type ($-1.91\,mV\,K^{-1}$). In addition, in the case of 50 pairs of p/n-type I^-/I_3^- redox couples connected in series, a voltage of 1 V can be obtained solely by harvesting body heat.

10.3.2 Functional Designs

Researchers can endow TECs with different characteristics by designing different electrolyte materials, allowing them to be flexibly applied in different scenarios. For example, the use of ionogel or hydrogel electrolyte in TECs system can effectively solve the problem of liquid leakage and enable TECs to work in a wider temperature range. In recent years, researchers have made great progress in the development of

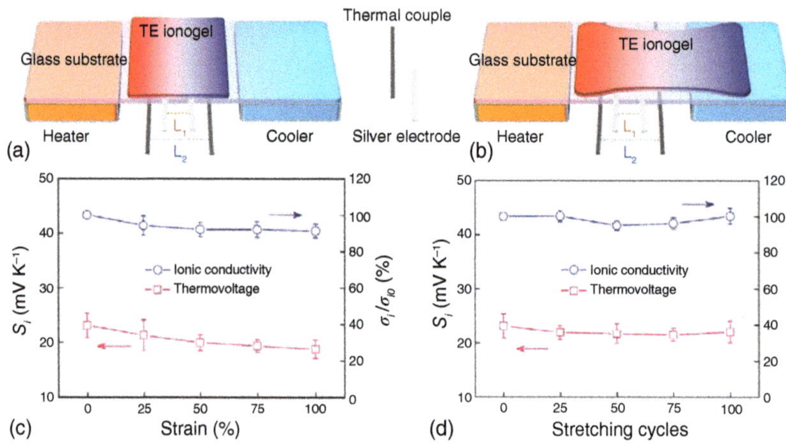

Figure 10.14 (a) Stretchable ionogel: Schematic diagrams of the stretchability. (b) Stretchable ionogel changes of thermopower and ionic conductivity of the ionogel under stretching. (c-d) Stretchable ionogel changes of thermopower and ionic conductivity of the ionogel under cyclic stretching. Source: Fang et al. [18]/John wiley & Sons.

ionogel or hydrogel electrolyte for TECs systems. Ouyang's team added ionic liquid EMIM: DCA into the designed elastomeric water-dispersible polyurethane (WPU) (containing both hard and soft segments) gels and developed a flexible and stretchable ionogel (Figure 10.14a) [18]. This flexible ionogel electrolyte can still maintain high TE properties and ionic conductivity under severe stretching (Figure 10.14b) or cyclic stretching (Figure 10.14c). By adjusting and optimizing the loading and of ionic liquids humidity, the *ZT* value (high Figure-of-merit) of this gel can reach to 1.3. Therefore, flexible, stretchable, and high-performance TECs are of great significance for the development of wearable self-powered devices.

In daily life, once a device is physically damaged, it cannot function properly. For inherently fragile inorganic materials, this has become a serious challenge. The good news is that the solid electrolyte based on self-healing polymers developed in recent years is expected to fully address this challenge. Jang's team reported a self-healing flexible TECs that uses anionic polyelectrolyte (poly(2-acrylamido-2-methyl-1-propanesulfonic acid)) (PAAMPSA), phytic acid (PA), and polyaniline (PANI) [19]. The hydrogen bonding and electrostatic interaction between PAAMPSA, PA, and PANI endow this material with self-healing properties. The hydrogen bonding and electrostatic interaction between PAAMPSA, PA, and PANI are key factors determining the self-healability ability of this electrolyte. As shown in Figure 10.15a, Malik's team added SiO_2 nanoparticles into the above-mentioned ternary TECs (Figure 10.15b). When this electrolyte undergoes a self-healing process, the thermalvoltage and ionic electrical conductivity of TECs can be fully restored to their previous state (Figure 10.15c) [20]. In addition, the Malik team's work has the highest *ZT* value (3.74) among currently known TECs and TEGs.

When the equipment is subjected to external forces and undergoes inevitable deformation, these developed flexible and self-healable TECs can better cope with

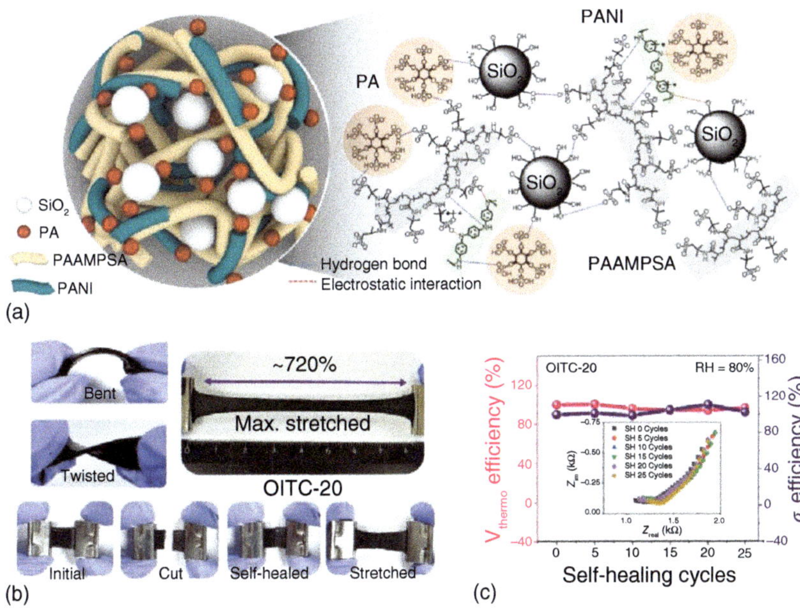

Figure 10.15 (a) Self-healable organic–inorganic hybrid film: schematic illustration of the molecular structure and interaction among components. (b) Self-healable organic–inorganic hybrid film: digital image illustration of the stretchability and self-healability. (c) Self-healable organic–inorganic hybrid film: thermopower and ionic conductivity efficiency of the film in self-healing cycles. Source: Malik et al. [20]/John wiley & Sons.

these situations. Recently, Wu's team designed a double-network hydrogel matrix electrolyte and developed a flexible TECs system based on it [21]. The carefully designed electrolytes not only solve the mechanical challenges but also effectively improve the power density of TECs. In addition, these TECs can maintain stable output power even under stretching [21].

Aqueous TECs face the problem of narrow operating temperatures, which is caused by the easy evaporation or freezing of aqueous solutions. To solve this problem, Gao's team developed an organohydrogel electrolyte that can be used in TECs system [22]. As shown in Figure 10.16, this electrolyte can work under low-temperature conditions (−20 centigrade). From the reported work, it can be seen that the design of multifunctional electrolytes has a significant impact on the energy conversion efficiency of TECs systems in TECs research. In addition, multifunctional flexible electrolytes have also expanded the application scenarios of TECs, which is of great significance for promoting the commercialization of TECs.

10.3.3 Device Integration and Applications

In recent years, increasing the energy density and efficiency of single cells has received increasing attention. However, for TECs, meeting the practical requirements still poses great challenges, mainly due to limited thermopower and

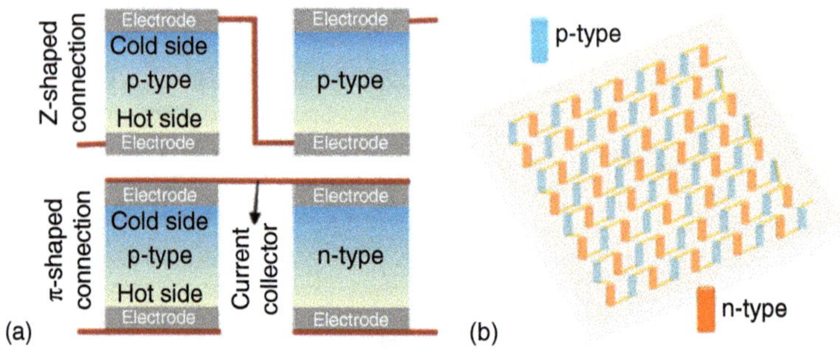

Figure 10.16 (a) Schematic diagram of the Z-shaped and P-shaped connection of single cells. (b) Schematic diagram of the connection of integrated thermoelectrochemical cells. Source: Liu et al. [23]/Royal Society of Chemistry.

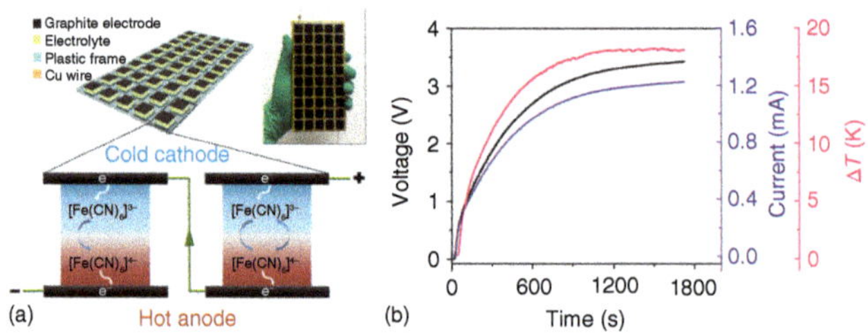

Figure 10.17 The application of the developed aqueous $Fe(CN)_6^{4-/3-}$ based thermogalvanic cell: the structure (a), the output performance (b). Source: Adapted from McCoy et al. [26].

temperature differences. In practical applications, in order to obtain effective thermal voltage, it is often necessary to integrate multiple single cells in a device [17, 24, 25]. There are two main options for connecting multiple single cells in series. The Z-shaped connection scheme shown in Figure 10.16a is only applicable to p-type TECs or n-type TECs. The P- shaped connection scheme is suitable for the mixed connection of p-type TECs and n-type TECs [23].

Figure 10.16b shows TEC systems that integrate multiple single cells into one device which can generate significant thermalvoltage and current for practical applications [23]. Figure 10.17 shows the developed $K_3Fe(CN)_6/K_4Fe(CN)_6$ aqueous TECs device, which integrates 50 single cells and can generate thermalvoltage of 3.4 V and current of 1.2 mA at the temperature difference of ~ 18 K [27]. In addition, the same group has developed another $K_3Fe(CN)_6/K_4Fe(CN)_6$ aqueous TECs device, which integrates 20 single cells and can generate a greater thermalvoltage of 3.1 V and current of 120 mA [27].

For TCCs, TE conversion applications are more complex. Figure 10.18 shows the energy conversion in the device which combines the capacitor and the hydrogel electrolyte, in which the electric energy comes from the ion diffusion and heat

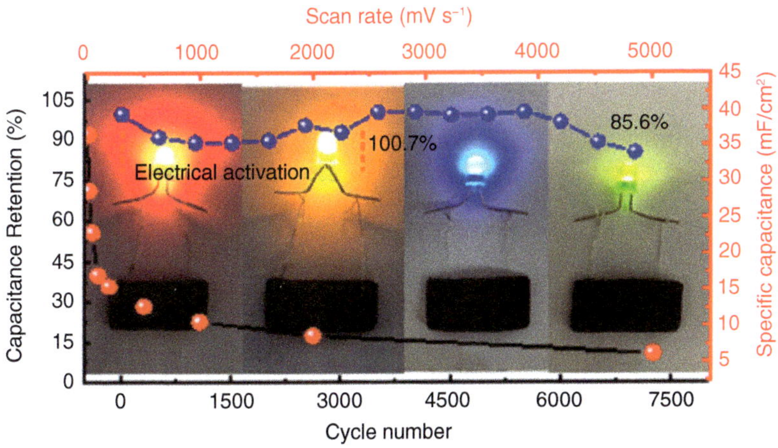

Figure 10.18 The developed NaCl-based thermally charged capacitors for heat to conversion by assembling with capacitors. Source: Yang et al. [28]/John Wiley & Sons.

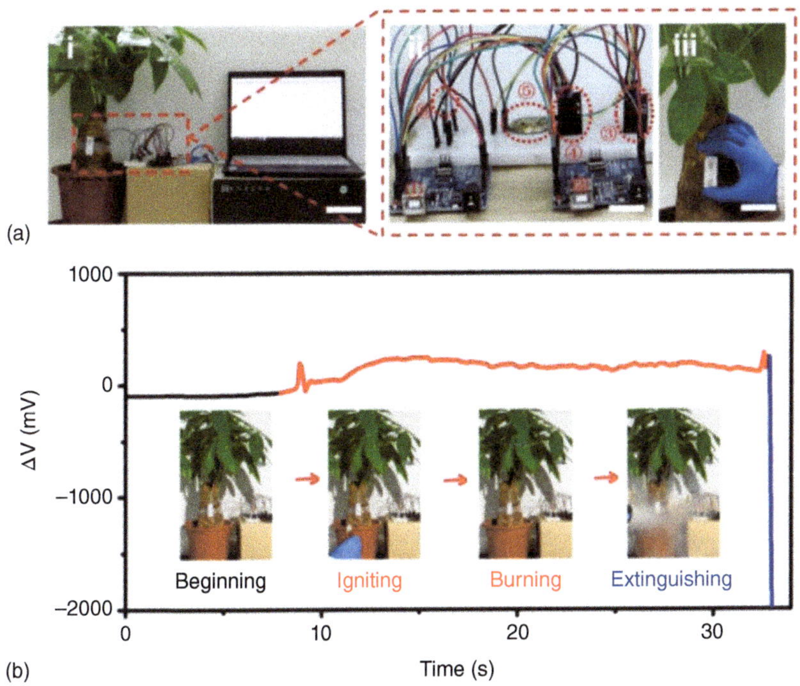

Figure 10.19 (a) Photographs of the wireless fire monitoring system. (b) Real-time evolution of thermoelectric voltage in the process of fire incident to the ornamental. Source: Wu et al. [29]/American Chemical Society

dissipation in the electrolyte [28]. However, due to the fact that ions do not pass through the external circuits, the application of TCCs device in the field of harvesting heat to electricity is severely hindered. In addition to generating electrical energy, thermal detection is also another important application field of TECs. In recent years, TCCs have been widely used in the field of thermal detection due to their enormous thermopower [15, 16]. As shown in Figure 10.19, based on the characteristics of TCCs, a fire early warning device has been successfully developed [29]. When fire occurs, the ambient temperature rapidly rises and thermal voltage is generated in TCCs and triggering an alarm.

10.4 Summary and Outlooks

There are many kinds of materials and devices in the field of TE chemistry field. From material to device, great breakthroughs have been made in Seebeck coefficient and *ZT* through protracted and unremitting efforts. However, finding materials with the added advantage of low thermal conductivity while at the same time having high conductivity remains a major challenge. How to achieve high *ZT*, we should not just concentrate on improving the power factor. While optimizing the power factor, we should pay attention to the structure-property dependence. There are many strategies, such as adding inorganic materials or carbon nanostructures as fillers, controlling doping, and enhancing the orderiness and crystallization of polymer substrates. These approaches need to be combined rather than applied separately. The development of high-efficiency polymer-based TE devices is still in its infancy, and the challenges posed by complex structure-property relationships need to be overcome. However, their inherent characteristics, such as low cost, flexibility, non-toxicity, solution workability, and other very promising research results, make us brave enough to explore their application in waste heat recovery technology. Substantial energy savings are no dream when collective approaches in the diverse fields of polymer materials and equipment manufacturing will enable us to make use of the waste heat that exists around us. Although in theory, thermal batteries can generate electricity from all sources of heat, their practical applications are very limited. A key problem is that waste heat, such as that from solar panels, data centers, and various electronic devices, is often produced as harmful by-products that mostly need to be dissipated. If the thermal cells are attached directly to these heat sources, their low thermal conductivity will inhibit heat dissipation and further cause the heat source to overheat. Therefore, typical static thermal cell devices are only suitable for collecting low-grade heat from sources that do not require heat dissipation, such as sunlight, the human body, and exhaust pipes. Alternatively, a flow heat battery may be an ideal choice that simultaneously removes and eliminates heat to keep the heat source within working range. Due to the high TE power of the thermocouple itself, the potential application of thermal cells in the field of temperature detection and cooling deserves further investigation. In the aspect of temperature detection, gel-based thermocouples as soft temperature sensors have broad application prospects in electronic skin, human activity monitoring, personal health care, and man-machine interface. Although the development of our electronic technology has made great progress in recent years, there are still some problems. The first is the application of polymers to gel substrates. In the interests of environmental protection and sustainable development, non-toxic and biocompatible polymers such as proteins, nucleic acids, and polysaccharides should be considered first. Additional packaging of the gel matrix is also something we need to consider, as it cannot be in direct contact with the human body, as well as the chemical and thermal stability of the REDOX pair when the TEC works at high temperatures for a long time. So, the study of TE chemistry is not just about the points that we seem to be thinking about.

With the continuous improvement of energy-saving requirements, the conversion and utilization of low-grade waste energy will gradually attract attention, so it is of great significance to manufacture a thermal-electric conversion device with good performance, commercialization, competitiveness, and high-cost performance. There are many potential utilization scenarios for heat charging capacitor devices that convert and utilize low-grade waste heat, such as large-scale data centers, automotive waste heat, geothermal, biomass heat, and ocean current heat. However, at present, the thermal-electric conversion experiment of heat charging capacitor is still in the verification and exploration stage, and there is still a long way to go from practical use. The research of these new systems still needs to continue to explore and discover new materials, new strategies, and new theories, and there is still a lot of work to be carried out.

(1) In terms of electrodes, the emergence of a new series of advanced nanomaterials with excellent electrical, thermal, and chemical characteristics, reasonable design of conductive electrode materials with layered and interconnected porous structures and REDOX characteristics is expected to improve the performance of heat-electric conversion and storage by providing additional pseudo-capacitance charge for the equipment.
(2) With regard to electrolyte advances, there is an urgent need to develop REDOX active electrolytes that are stable in a wider potential window, with high ion and low thermal conductivity, while being eco-friendly, cost-effective, and safe.
(3) Since the overall efficiency of the system depends on energy conversion and storage processes, the role of different device components (electrode and electrolyte materials, diaphragm, etc.) should not be analyzed in isolation, but in an integrated manner during collection and storage. In order to narrow the gap between conversion and storage efficiency and improve overall performance, a thorough understanding of the possible interactions and synergies between the different components of the device should be made. In particular, it is critical to ensure an ideal match between electrode material (e.g. composition, morphology,

pore structure, etc.) and electrolyte (ionic conductivity, electrolyte ion size, and electrochemical stability) properties, as well as to improve electrode/electrolyte interface contact, while determining electrolyte migration rate under temperature gradient.

(4) Because electrode performance is highly temperature dependent, i.e. their electronic structure, conductivity, work function, and the type and intensity of interfacial interactions with electrolyte ions, the charge storage contribution of the electrodes under temperature gradients needs to be considered. In addition, the temperature dependence of reactions that may occur in the electrolyte and at the electrode/electrolyte interface (in the case of redox-active electrodes) should be investigated in depth, as they can help to adjust the TE capacity characteristics of the device in conjunction with the Soret effect.

(5) Complementary in situ characterization of devices is essential for an in-depth understanding of the fundamental mechanisms involved in thermal charging and storage processes, allowing real-time tracking of local structural and electronic changes occurring on electrode surfaces/interfaces. At the same time, theoretical simulation and modeling are essential tools for optimizing equipment configuration and improving its performance at the electrode and electrolyte levels. These studies help clarify the synergies between energy harvesting and storage components and predict the output of the device.

References

1 Zhu, T., Liu, Y., Fu, C. et al. (2017). *Adv. Mater.* 29: 1605884.
2 Güler, N.F. and Ahiska, R. (2002). *Appl. Therm. Eng.* 22: 1271.
3 Li, Z., Xu, Y., Cui, J. et al. (2023). *J. Power Sources* 555: 232386.
4 Li, Z., Xu, Y., Wu, L. et al. (2022). *J. Mater. Chem. A* 10: 21446.
5 Li, Z., Xu, Y., Wu, L. et al. (2022). *Nat. Commun.* 13: 132.
6 Xu, Y., Li, Z., Wu, L. et al. (2023). *Batter. Supercaps* 6: e202200331.
7 Kobayashi, W., Kinoshita, A., and Moritomo, Y. (2015). *Appl. Phys. Lett.* 107: 2.
8 Meng, T., Xuan, Y., and Zhang, X. (2021). *ACS Appl. Energy Mater.* 4: 6055.
9 Cheng, C., Wang, S., Tan, P. et al. (2021). *ACS Energy Lett.* 6: 329.
10 Gao, C., Lee, S.W., and Yang, Y. (2017). *ACS Energy Lett.* 2: 2326.
11 Zhang, H. and Wang, Q. (2021). *Chem. Phys. Rev.* 2: 021304.
12 Lee, S.W., Yang, Y., Lee, H.W. et al. (2014). *Nat. Commun.* 5: 3942.
13 Wang, X., Huang, Y.T., Liu, C. et al. (2019). *Nat. Commun.* 10: 4151.
14 Liu, S., Yang, Y., Huang, H. et al. (2022). *Sci. Adv.* 8: eabj3019.
15 Zhao, D., Martinelli, A., Willfahrt, A. et al. (2019). *Nat. Commun.* 10: 1093.
16 Manigrasso, J., Chillón, I., Genna, V. et al. (2022). *Nat. Commun.* 13: 41467.
17 Duan, J., Yu, B., Liu, K. et al. (2019). *Nano Energy* 57: 473.
18 Fang, Y., Cheng, H., He, H. et al. (2020). *Adv. Funct. Mater.* 30: 2004699.
19 Akbar, Z.A., Jeon, J.W., and Jang, S.Y. (2020). *Energy Environ. Sci.* 13: 2915.
20 Malik, Y.T., Akbar, Z.A., Seo, J.Y. et al. (2022). *Adv. Energy Mater.* 12: 2103070.
21 Lei, Z., Gao, W., and Wu, P. (2021). *Joule* 5: 2211.

22 Gao, W., Lei, Z., Zhang, C. et al. (2021). *Adv. Funct. Mater.* 31: 2104071.
23 Liu, Y., Cui, M., Ling, W. et al. (2022). *Energy Environ. Sci.* 15: 3670.
24 Yang, P., Liu, K., Chen, Q. et al. (2016). *Angew. Chem. Int. Ed.* 55: 12050.
25 Zong, Y., Li, H., Li, X. et al. (2022). *Chem. Eng. J.* 433: 134550.
26 McCoy, D.E., Feo, T., Harvey, T.A., and Prum, R.O. (2018). *Nat. Commun.* 9: 1.
27 Yu, B., Duan, J., Cong, H. et al. (2020). *Science* 370: 342.
28 Yang, X., Tian, Y., Wu, B. et al. (2022). *Energy Environ. Mater.* 5: 954.
29 Wu, X., Gao, N., Zheng, X. et al. (2020). *ACS Appl. Mater. Interfaces* 12: 27691.

Index

Low-Grade Heat Harvesting: Materials, Devices, and Technologies, First Edition. Xiaogang Zhang.